HOUSING PHILOSOPHY

This book addresses key issues in housing policy through the lens of the philosophical concepts that underpin them. It is intended to be an introduction to philosophical subjects and how they relate to key housing issues and deals with the concepts with enough rigour and depth to be of use to undergraduate and postgraduate students and teachers.

Both practitioners in the housing sector and academics researching housing-related subjects often use terms such as 'fairness', 'social justice', 'tenure', 'property', 'home' and others as if there is a single agreed meaning for these terms. But these terms can be highly contested, and there are multiple viewpoints for each of them that could change how we approach them, and how we therefore create, interpret and implement policies and procedures. This book aims to introduce certain concepts and provide guidance and stimulate thinking around how they make an impact on real-world policy. Each section opens with a relevant case study designed to highlight the philosophical concepts to be discussed and coverage ranges from homelessness to the role of government and the state, to house prices, value and property rights.

This book will be of interest to students at both undergraduate and postgraduate levels, studying housing qualifications or philosophy, sociology, politics, governance and social policy – it will be of particular use for those interested in the application of philosophical concepts to a real-world policy area with clear consequences.

Yoric Irving-Clarke has worked in housing for over 20 years and is the author of *Supported Housing: Past, Present and Future* and co-author of *Housing and Domestic Abuse: Policy into Practice* with Dr Kelly Henderson. He has a PhD from De Montfort University, UK.

HOUSING PHILOSOPHY

Applying Concepts to Policy

Yoric Irving-Clarke

Routledge
Taylor & Francis Group
LONDON AND NEW YORK

First published 2023
by Routledge
4 Park Square, Milton Park, Abingdon, Oxon OX14 4RN

and by Routledge
605 Third Avenue, New York, NY 10158

Routledge is an imprint of the Taylor & Francis Group, an informa business

British Library Cataloguing-in-Publication Data
A catalogue record for this book is available from the British Library

ISBN: 978-1-032-27818-8 (hbk)
ISBN: 978-1-032-27606-9 (pbk)
ISBN: 978-1-003-29821-2 (ebk)

DOI: 10.1201/9781003298212

Typeset in Bembo
by codeMantra

CONTENTS

ACKNOWLEDGEMENT

Once again, thanks to Dr Peter King for his, as ever, generous support with his time and perceptive comments on draft chapters. And, thanks to all those who have wittingly or unwittingly contributed to this book via purposeful or random conversations. Finally, thank you to Cheri, Tash and my family for their love and support with all that I do. These are simply debts I will never be able to repay.

PROLOGUE

In this book, and in my other writings, I have constructed a history of housing policy in the UK and used this at various points to illustrate the points I make. To provide you, the reader, with a quick reference point and to provide a full account in one place, I have pulled that history together here.

A brief history of housing policy in the UK

(Social) housing policy has a relatively short history in the UK with much of it being constructed as part of the post-war welfare state (c.1945–1979). Prior to this housing for the poorest people was via Houses of Industry or Correction, workhouses or 'lunatic asylums' (See Irving-Clarke [2016; 2019] for a full account). There were also pockets of charitably provided housing, for example, alms-houses provided by religious groups or the mercantile class (See Pannell & Thomas, 1999). Otherwise, housing was largely a mix of owner occupiers and private renters. Many people lived in accommodation that was tied to their jobs, e.g. housing provided by landowners for farm staff or later in villages such as those pioneered at Bournville by the Cadbury family. The Industrial Revolution had meant forced clearances of much common land and people being forced into the new cities to man the factories.

That said, the *Housing and Working Classes Act 1890* empowered local authorities in London to build housing for poorer communities and charge reasonable, fair rents. This was the first time the state had played a role in the provision of housing to a noticeable extent. Councils had to re-house a least half of the people displaced by these slum clearances. A decade later, the *Housing of the Working Classes Act 1900* extended the 1890 Act to areas outside London and allowed councils to buy land outside their own districts.

Despite these advances, at the start of World War I, nearly 90% of households lived in the private rented sector, with most of the remainder being owner occupiers (King, 2006). The outbreak of the war led to large-scale demographic changes as working-age men were conscripted to fight, leaving gaps in the labour force. The need to produce munitions and other war supplies meant that those left (mainly women) moved *en masse* to cities to work in factories (in an echo of the Industrial Revolution). Construction of new homes was also at a standstill, as materials were directed to the war effort – this led to a shortage of urban housing, and as a result rents increased dramatically. This led to 'rent strikes' in areas such as Glasgow (whose shipyards were vital to the war effort), in response the government introduced frozen rents and mortgage interest at 1914 levels via the *Rent and Mortgage Interests (War Restrictions Act) 1915*. The housing shortage was only made worse after the war ended as soldiers returned and started families (King, 2006). This led to the roll out of the subsequent 'Homes fit for Heroes' (Yorke, 2017) via the *Housing and Town Planning Act 1919* (also known as the Addison Act) which introduced a national housing programme that highlighted the shortage of good quality homes for men returning from the war.

The origins of social housing, as we think of it today, are commonly traced back to the early 20th century and Octavia Hill's efforts around the London slums. Her overall aim was to make "lives noble, homes happy and family life good" (www.octaviahill.org). Her moralistic and paternalistic approach was that properties should be let to tenants at a high standard and in return, tenants would be expected to 'moderate' (change) their behaviour. This approach is still in evidence today in both housing and welfare policy. For example, Carr et al. (2007) see crime control in relation to housing as a new problematisation of the housing crisis, adding that this has echoes of the foundations of housing policy in the 19th century and Victorian concerns around pauperised deviance (Cowan & McDermont, 2006). Using policy in this way has a long history, see Irving-Clarke (2016; 2019). Burnett (1993) argues these factors were also pivotal in terms of post-war social policy.

The interwar years (1918–1939) saw an explosion in owner-occupation as wages rose in real terms, house prices remained static and the creation of building societies with high levels of liquidity allowed a stable and secure base for investment. Thus, the sector with the highest growth between the wars was owner-occupiers (King, 2006). There was, however, an acceptance for the first time that private enterprise would not be able to produce enough quality homes at affordable prices and that the state had a role. The idea that the market could not provide homes for all would lead to a 'safety net' provided by the state to provide for those who were unable to access the market.

World War II had a dramatic effect on the provision of housing by the state. The government was faced with the two-pronged problem of a housing shortage (caused by bomb damage) and, returning soldiers and the related "baby-boom" and growth in household formation (Irving-Clarke, 2016). The Beveridge Report (1942) introduced the concept of a comprehensive state safety net to protect its citizens from the five giant evils in society (squalor, ignorance, want, idleness, and disease) from cradle to grave. The Labour government, led by Clement Atlee, chose to meet the demand for new homes mainly via large-scale local authority building programmes. House building numbers remained at over 150,000 units per annum right through the 1950s and were still above 80,000 per annum throughout the 1970s. Council housing was in good financial shape with many older properties free of debt. The income from these homes could therefore be used to enable reductions in rent or subsidise further building (King, 2006).

By 1971, councils housed almost one in three households (and would do so until the early 1980s), owner occupation was also growing and fewer than one in five households were privately renting – this number would continue to fall. In addition, two-thirds of working-age social-renters were in full-time jobs, the broad tenant base and strong position of social housing could have been a springboard for further expansion (Stephens et al., 2018). However, Tucker (1966) argues that the high-volume building for slum clearance did not have the earlier post-war commitment on building quality homes resulting in both the standard and status of social housing falling. At the same time, the political link between housing and its impact on all areas of life was established. Housing was promoted as a crucial political issue; the manifesto on which the Conservatives fought the 1951 election, which stated:

> Housing is the first of the social services. It is also one of the keys to increased productivity. Work, family life, health and education are all undermined by crowded houses. Therefore, a Conservative and Unionist Government will give housing a priority second only to national defence.
>
> *(Conservative Party, 1951)*

Despite seeing housing as second only to national defence at the 1951 General Election, the Conservative government of Harold Macmillan (1957–1963) was the first to challenge the 'post-war consensus' by questioning the mass provision of housing by the state. However, housing is often cited as 'the wobbly pillar' under the welfare state (Torgerson, 1987, pp. 116–126), because of the partial extent in which it has been considered as a responsibility of government and of social policy.

The ideological debate about the state's role in mass housing provision and the encouragement of those who could do so to move out of social housing was twinned with the fact that social housing was becoming less easily available. Local authorities were encouraged to allocate council housing on *need* rather than desert following the Cullingworth Report (1969) meaning that the essence of social housing was changing. As Mullins and Murie (2006) point out, council housing had always been provided to those who could afford the rent, e.g. the more affluent working class and that those who could not afford to pay were residualised in the sub-standard private rented sector (or worse). In 1979, 20%

of families in the top decile of income distribution lived in social housing, compared to almost zero by 2004/05 (Mullins & Murie, 2006).

Prior to the 1970s, council rents were directly linked to the actual cost of provision (as opposed to being market driven); there were no "subject subsidies", e.g. compulsory rent rebates or housing benefit to assist with housing costs (King, 2006). These were not introduced until the early 1970s (King, 2006). The introduction of compulsory rent rebates in the *Housing Finance Act 1972* broke the link between cost of provision and rents charged and began the residualisation of the poorest people in council housing by removing the cost of a social rent as a disqualifying factor. Residualisation is defined by Malpass and Murie (1982, p.174) as follows:

> the process whereby public housing [and other social housing] moves towards a position in which it provides only a 'safety net' for those who for reasons of poverty, age or infirmity cannot obtain suitable accommodation in the private sector.

Until this point, council housing had been an aspirational tenure for the relatively wealthy working class (CIH, 2018). Prior to the 1972 Act, local authorities did have the power to charge a subsidised rent on a discretionary basis but as Cowan and Marsh (2018) point out, the lack of security of tenure (only introduced in Right to Buy legislation during the 1980s) left households vulnerable to eviction at the whim of local authorities; and withdrawal of subsidised rent could be used as a lever to produce a desired result, e.g. moving to a smaller property in cases of under-occupation (CHAC, 1953, pp. 14). Although Cowan and Marsh note that this was not common practice.

The *Housing (Homeless Persons) Act 1977* continued the residualisation of the poorest people in council housing; local authorities now had a duty to house people considered vulnerable under the law and provide financial assistance to help them afford the rents. Council housing was thus used to discharge this duty and the introduction of compulsory rent rebates (subsequently Housing Benefit and now the housing element in Universal Credit) meant that, as stated previously, affordability was no longer a disqualifying factor; although Malpass and Victory (2010) find that this residualisation was implicit in and result of housing policy throughout the 20th century.

Although some argue that the neo-liberalisation of welfare policy has been in train for some time (see below), the election of a transformed Conservative Party under Prime Minister Margaret Thatcher in 1979 and her adherence to the monetarist policies of Milton Friedman, which prescribed limited state intervention in a 'self-correcting' free market, drove housing policy during the 1980s and well beyond (Irving-Clarke, 2016; 2019). Although, Gough (1980) suggests that the ideological attack on the post-war welfare state was largely initiated by the Labour governments of Wilson and Callaghan. He further argues that the qualitative shifts in social policy at this time were aimed at reasserting individualism, promoting self-reliance and family responsibility and ultimately dismantling the ideology of the collective social provision of the post-war era. The Conservative government focused on setting people free from government constraints on their lives by rolling back the state, re-emphasising a commitment to market provision of housing, a revival of the private rented sector and a belief in extending home ownership with council (social) housing being aimed only at those who could not compete in the market.

The emergence of neo-liberal policy represented a sea change in ideological approach to welfare and housing (Gough, 1980). Driving policy was a preference for home ownership over other tenures (Gurney, 1999) depicted in the *Housing Act 1980* which introduced the flagship 'Right to Buy' policy. Local authority tenants could now purchase their property at a discount dependent on how long they had been a tenant. The policy proved extremely popular with voters and has never been fully repealed by successive governments of either party in England; although the *Housing Act 2004* introduced stricter regulations, including a tighter time frame in which tenants could sell after purchase and with less discount to buy (King, 2006). The Scottish and Welsh governments have both recently ended it in a bid to protect social housing stock from further reductions, and to ensure that the state can provide safe, secure accommodation to those who are unable to buy or rent their own home (Stephens, 2019; Irving-Clarke & Henderson, 2021).

The Conservative governments of the 1980s and 1990s continued the doctrinaire repositioning of state-controlled housing out of the government arena via the *Housing and Planning Act 1988* which gave councils the option of transferring all or part of their housing to another landlord, such as a Registered Social Landlord (RSL). The 1988 Act also allowed council tenants (as a whole) to choose to transfer their existing homes to another landlord, either a separate housing association (large-scale voluntary transfer, LSVT) or arms-length management organisation (ALMO) operated by the local authority but managed autonomously.

The effects of the 'Right to Buy' policy were manifold and widespread. It removed much of the most desirable and sought-after stock from the social housing sector, without a replacement-building programme (Forrest & Murie, 1988). This point is borne out by Manns (2017) who stated that since the introduction of the policy, it is estimated that over two million properties have been moved into private ownership. It also served to deepen the divide between owner occupation and the social rented sector. Lund (2011, pp. 142) argues that Right to Buy and other policies resulted in a continued concentration of low-income households in the social housing sector with those accessing social housing increasingly in the greatest housing and social need. Social housing was becoming more synonymous with welfare, and a clear polarisation of the housing market was established. Hills argues that housing policy hardly existed anymore but survived within the welfare state, "mainly as an adjunct of social security (through Housing Benefit) or as part of wider city regeneration policies" (Hills, 1998 pp. 13 quoted in Tunstall, 2015).

Housing policy (arguably) changed again with the election of New Labour under Tony Blair in 1997. Although they continued the policy of seeing home ownership at the pinnacle of the housing system, they also initiated a major investment in the social housing stock under the Decent Homes Programme (DETR, 2000), invested in housing-related support via the Supporting People programme (DTLR, 2001) and made major inroads into tackling homelessness and rough sleeping. However, the Thatcherite 'right to buy' policy continued as did the transfer of housing stock out of direct local authority control. These measures further residualised social housing as a tenure for those on the lowest incomes (CIH, 2018). Pawson and Mullins (2010) argue that New Labour perceived council housing as a 'redundant project' with low-demand, poor-quality stock and saw the issue through the lens of poverty and limited life chances and opportunities, not merely as a matter of giving tenants more choice of landlord (via ALMOs or LSVT [see below]). Labour recognised there was a stigma attached to the widening difference between homeowners and those living in estates with low demand and poverty, with the term 'sink' estate becoming used by many commentators (See Cm 4045). Although there was recognition that welfare and housing policies needed to be better aligned (Malpass & Victory, 2010).

However, New Labour did recognise the need to improve the housing product on offer. The housing Green Paper, "Quality and Choice: A Decent Home for All" (DETR, 2000) aimed to bring all social housing into decent condition by 2010. Local housing authorities with remaining stock were required to undertake an options appraisal to identify how they would secure resources to meet the new Decent Homes Standard (DETR, 2000). For many, the only option to raise funds to improve homes was to consider an LSVT. This continued the Conservative approach of removing housing from the public sector with almost a million homes removed from local authority ownership via LSVT between 1997 and 2009; reflecting the Conservative (neoliberal) approach of the state in an enabling role, in relation to housing, as opposed to providing and managing it.

Because of its pretensions to present a route between neo-liberalism and social(ism) democracy, the third way has been criticised as being an ideological vacuum, little more than pragmatism in place ideology. It has been asked whether the Third Way is really a radical programme of renewal or a clever marketing gimmick for a continuation of Thatcherism (King, 2001) – neoliberalism with a human face.

The Grenfell Tower fire and its aftermath

Following the financial crisis of 2008, the Conservative Party was returned to power in coalition with the Liberal Democrats. Despite maintaining the primacy of home ownership as a policy goal, the coalition government embarked on possibly the biggest shake-up of the welfare system since its inception. The changes they made are covered fully in Chapter 6 but are summarised here.

Six means-tested legacy benefits were combined into Universal Credit (UC). Local Housing Allowance (LHA) rates were reduced to paying the 30th percentile rent amounts in local housing markets from the previous 50th percentile, thus shrinking the number and quality of homes available to rent in the private sector. LHA rates were also frozen at this level, meaning that they were further eroded by rent inflation for several years. The 'bedroom tax' was introduced to encourage households in houses with too many rooms for their assessed needs to downsize. However, this was subject to several successful court challenges, and there were often no properties to downsize too or removal costs prohibitive. The 'benefit cap' limited the amount in benefits households could receive in welfare. The welfare system, in general, was made more conditional and a significantly more punitive sanctions regime introduced.

Two major scandals in the social housing sector placed focus on the sector and forced the now Conservative only government to act.

The standard of social housing was once again thrown into the spotlight in 2017 by the tragic fire at Grenfell Tower in North London. The fire highlighted not only that the tower block had been covered in dangerously flammable cladding but also that appropriate checks had not been carried out, fire doors and other essential safety equipment was missing and the 'stay put' policy was woefully inadequate. Later in around 2021, ITV news uncovered widespread disrepair on the Eastfields Estate, also in London and owned and managed by a social landlord, Clarion Housing Group. At Grenfell Tower, the residents had raised a string of concerns about fire, and wider, safety in the block; the tenants on the Eastfields Estate had also repeatedly reported disrepair in their homes. These issues raised serious questions about the way that social tenants are treated both by their landlords and more broadly. The stigmatisation of social housing and social housing tenants has been an issue in the sector for some years, with campaign organisations such as See the Person and the Chartered Institute of Housing taking up the cause in recent years. In response, the Conservative government published both a Social Housing Green Paper and a Social Housing White Paper that promised to both improve conditions in social housing and the way that landlords and the wider public view social housing tenants.

A new tripartite regulatory arrangement for social housing is now in place. The Regulator for Social Housing (RSH) has published new, stronger principles it will apply in regulating the sector, and the draft outcomes it will be assessing for. Once the government passes the 'Social Housing Regulation Act', giving shape to the role of the regulator, there will be further consultations on the regulatory standards and how the inspection regime will operate. RSH is already consulting on the proposed 'Tenant Satisfaction Measures' that will be used to assess landlords. At the time of writing, we are awaiting the publication of these measures and their passage through Parliament.

On building safety, the *Building Safety Act 2022* had recently completed its passage through Parliament at the time of writing. In summary, the bill creates a clear, proportionate framework for the design, construction, and management of safer, high-quality homes, and strengthens the regulatory regime around construction products to ensure they are safe. Building owners will be required to manage safety risks with clear lines of accountability for safety during design, construction, completion and occupation. Those who don't meet their obligations under the proposed Act may face criminal prosecution.

The Housing Ombudsman Service is already largely operating under its new remit, publishing a wide variety of reports, data and training materials. More broadly, the Future Homes Standard continues to shape environmental standards for new homes, with the building regulations being revised to support its goals. An updated Decent Homes Standard, proposals for electrical safety checks in social housing, vehicle charging points in new homes and greater protections for private renters are also forthcoming.

Embedded throughout the Social Housing White Paper, professionalisation and stigma overarch many of the other themes. The government has convened a working group to review the training and staff development of social housing staff. The Chartered Institute of Housing have launched a professional standards framework (REF) for people working in housing, and it seems likely that this may form a key part of any new approach to professional development coming out of the working group.

At the time of writing, we are waiting for detail on further reforms. The regulations from the *Building Safety Act 2022* will shortly be laid before Parliament, as will the full Social Renters' Reform Bill. The RSH has recently consulted on a number of 'Tenant Satisfaction Measures', including several perception measures asking how tenants feel about their property and their landlord. This is a likely new direction of the regulatory framework. The full framework and how it will be implemented is yet to be published as elements of it are dependent on the Social Housing Regulation Bill, mentioned above. We do know that there is likely to be regular inspections of landlords with over 1,000 units; however, following recent hearings of the Department of Levelling Up, Housing and Communities Select Committee, it seems unlikely that these inspections will include site visits, except where there are serious concerns. This is a major departure from the previous inspection regime of the Audit Commission that included a rigorous process with both site visits and requirements to provide evidence of compliance with 'Key Lines Of Enquiry' (KLOEs).

The impact of these changes will be profound and widespread.

INTRODUCTION

This book has its genesis in many discussions with family, friends and colleagues about housing and the big ideas involved in its provision and use. I am fortunate in that my life encompasses both housing policy and practice and the academy, this means that I get to read and hear ideas about both sides of what is a two-sided coin. I use that analogy advisedly. All too often, housing policy and practice, and philosophy face different ways and never the twain shall meet; almost Janus faced. I do not wish to suggest that one looks back whilst the other looks forward; merely that they rarely engage with one another face to face. There are, of course, exceptions. Esteemed books by colleagues (and I am pleased to say friends), Helen Taylor (2018), Beth Watts and Suzanne Fitzpatrick (2018), and of course my erstwhile PhD supervisor Peter King (2001; 2003; 2004; 2011, 2016) have applied philosophical concepts to housing. Indeed I, myself, did so in my first book (Irving-Clarke, 2019).

However, working in the housing sector and particularly for the Chartered Institute of Housing, I have been struck by the way people in the sector cling to certainties and definitions of terms, without an understanding of the arguments or concepts that undergird them and the broader social arrangements in the country generally. I do not, of course, wish to imply that people in the sector are in any way dogmatic or inflexible; but that there is a philosophy-shaped gap. Perhaps given the open paragraph, this should not be a surprise. For many, housing is a practical arena not suitable for theorising, as King (2011) asks, who has time for theory when there are people sleeping on the streets?

As a specific example, it is common to hear people within the housing sector referring to 'fairness' or 'justice' as if we all agree what they mean. Many in the sector probably do agree the meaning of the word when they use it. But justice for whom? Fairness for whom? How is justice or fairness defined? How do we know

DOI: 10.1201/9781003298212-1

when it has been achieved? Raising taxes to build more homes for those who are not adequately housed, but what impact might that have on households who are struggling to survive but have their taxes raised? Even if the government only taxes the wealthy, how does it decide who the wealthy are? How much should they tax them? Do not the wealthy have property rights, and the right to keep the money they work to earn or the wealth they own, regardless of what 'good' others think they could do with it? Is there a substantive difference between some types of wealthy people compared to others? For example, is it more acceptable to tax someone who has largely inherited their wealth, when compared to people like Sir Alan Sugar who came from a relatively modest background but is now a billionaire thanks to his hard work and business acumen?

Where the government owns and manages its own housing or funds the building of social housing, who gets to live in it? Given that housing, along with pretty much everything else, is a scarce resource, how should it be allocated? What should be taken into account in making these decisions? Should people who have made themselves 'intentionally' homeless, by not paying the rent or committing anti-social behaviour, be excluded from being housed by the state? Don't they still need somewhere to live, after all?

Whilst I cannot promise that I will come to definitive answers to these questions, for that is not my purpose, what I do want to do is provoke readers' thoughts about how we make these decisions, or at least how we think about the problems.

Each chapter takes one concept from the housing literature and attempts to strip it back to its basis in theory or concept. I also attempt to discuss how each concept has developed over time, drawing on examples from housing and other policy areas where it is relevant. Each chapter starts with a question relevant to the subject in hand. My purpose here is not to be controversial or particularly provocative, but I do want to kick-start and encourage discussion of the topic at hand. My intention in writing this book is that it is used as part of housing courses and in the workplace, to encourage thought and discussion. The structure of the book is as follows.

The opening chapter deals with ideology. These are the big political ideas that underpin many of our ideas and opinions about the world and different political parties' and governments' proposed policy solutions. I cover the main ideologies present in the UK political system Conservatism, Socialism/Communism, Liberalism and Neo-Liberalism along with examples of how these have translated into policy ideas and prescriptions.

In Chapter 2, I look at the state, the theoretical justifications for its existence and ideas about the size of the state and what its functions should be. Starting from first principles, I attempt to show why we need the state at all. The chapter also includes some comparisons with the US to bring clarity to the discussion and show a different view but from a country with a relatively similar system.

Chapter 3 looks at markets. Again, here I begin the chapter taking the discussion back to first principles and how markets, the use of money and the development of the banking system. Housing markets are distinct in several ways and have their own features and peculiarities. The chapter looks at these with a view ascertaining why houses in London are so much more expensive than the rest of the country.

Chapter 4 looks at homelessness policy and issues of social justice. I look at the various definitions of social justice that have been given and are generally accepted, before moving onto discuss social justice and housing, and how housing fits into the various conceptual frameworks given for the concept of social justice.

Chapter 5 looks at property rights and how ownership of private property can be justified, manifested, and maintained. The chapter looks at property rights from the distinct ideological viewpoints given in Chapter 1, and links with concepts from the chapters on economics and social justice.

Chapter 6 looks at policy change. It is often said that one of the strengths in the British system is that it produces governments with a distinct programme for government, as enumerated in their manifesto, a strong mandate for that programme; and a majority in Parliament so it can actually govern successfully. Given that this is the case, can a government do exactly as it pleases with its mandate and majority? Or are there some clear, and perhaps not so clear barriers in its way? This chapter looks at some of these barriers and how and why policies might change.

Finally, the issues I have chosen to discuss are not mutually exclusive and do intersect considerably. I there close the book with a concluding chapter that attempts to bring some of these themes together.

References

Irving-Clarke, Y (2019) *Supported Housing: Past, Present and Future*, Routledge, Abingdon.

King, P (2001) *Understanding Housing Finance*, Routledge, Abingdon.

King, P (2003) Housing as a freedom right, *Housing Studies* 18(5) pp.661–672.

King, P (2004) *Private Dwelling: Contemplating the Use of Housing*, Routledge, Abingdon.

King, P (2011) Using Big Ideas: The application of political philosophy in housing research, *Housing, Theory and Society* 28(2) pp.109–122.

King, P (2016) *Principles of Housing, Routledge*, Abingdon.

Taylor, H (2018) *Social Justice in Contemporary Housing*, Routledge, Abingdon.

Watts, B & Fitzpatrick, S (2018) *Welfare Conditionality*, Routledge, Abingdon.

1

CAN'T WE ALL JUST AGREE THAT HOUSING IS A GOOD THING, AND THE GOVERNMENT SHOULD PROVIDE IT?

Case study

Nearly 300,000 households (many more individuals) are either already homeless or in danger of becoming homeless within weeks, in England. In this context, homelessness is not only sleeping on the streets, although we know this is the most damaging form of homelessness. It can also mean a household is living in sub-standard accommodation that may be damp, mouldy or in poor repair. Their housing many not be large enough for the number of people living there (overcrowded), or one or more persons in the household may be at risk of harassment and/or abuse.

Further, there is not enough good quality, affordable housing to ensure that everyone has a suitable place to live. There is a solution to this. The government could simply use its powers to seize land and build and manage enough housing to make sure that everyone has a home that suits them. This has the added advantage that all revenue from this housing would go back to the government and be used to pay for the construction costs, and for more new housing to be built.

We know that rough sleeping is very harmful to health and that the harm is cumulative – the longer you sleep rough, the greater and longer term the harm is.

We also know that being well-housed is one of the best things for a person's health and wellbeing across all other areas of their life.

DOI: 10.1201/9781003298212-2

Questions

- Why can't we just agree that the state should do this?
- What might the downsides to this course of action be? Are there any?
- What rights might people have that either support or negate this as a solution?

We all need somewhere to call home. Home is a merit good (King, 2016); it is a good thing in and of itself. It needs no other justification. So, if that is the case, why cannot we all just agree that the government should provide it. We can all pay rent to the government, once the government has paid the cost of building the initial tranche of homes, it can spend the proceeds on more and more housing, until everyone in the country lives in government-owned housing paying rent to the state. Indeed, in the middle part of the 20th century, much of the government-owned (council) housing had indeed paid for itself and the government was able to use the proceeds to construct more (King, 1996). If they had carried on, everyone could now be living in a government-owned, high-quality, low-rent property. So why aren't we?

Well, perhaps if the process described above had continued, we would be. But they famously did not. The election of a Conservative government in 1979 under the leadership of Margaret Thatcher ended any chance of that. Thatcher and her colleagues distrusted the state and in particular local authorities, rolling back the role of the state in people's lives was a key goal of her administration. And, to some degree with good reason.

Industry and workers found themselves at almost permanent loggerheads. Strikes and other industrial action were commonplace, there was conflict as we fought one another for a share of the ever-shrinking economic cake. Britain found itself unable to innovate, trade and prosper and gradually became "the sick man of Europe" (Blond, 2010, p. 94), culminating in the 'winter of discontent' in 1978/1979. Between 1950 and 1973, average gross domestic product (GDP) growth in the UK was only 2.5%, whilst the western European average was 4%; Germany and Italy both averaged 5%. These countries managed to eliminate around 40% of the productivity gap between them and the US, with France at 74%, Italy 78% and Germany at 79% of US productivity, measured by hours worked as a percentage of GDP. Exports fared little better with the UK annual export growth rate between 1950 and 1973 at only 6.9%. This can be compared to an EU average export growth rate of 12.2%, and a remarkable German export growth rate of 19.8% (Eichengreen, 2007).

There was a "false Keynesian consensus" (Blond, 2010, p. 102) in the UK, between owners who were only interested in higher personal dividends and workers who were only interested in increasing wages for work of reducing quality and efficiency. Both retreated into their comfort zones. Management became diffident, un-visionary and lazy, choosing comfort over innovation and challenging workers. The labour unions (workers) retreated into opposing anything that seemed to challenge their narrow, sectional interests and 'class struggle'.

Managerial incompetence, coupled with worker intransigence, segueing into ideological hostility to capitalism itself.

Aside from these practical factors, Thatcher had an ideological distrust of government and state spending and services. She believed in a much smaller state apparatus that intervened much less in the lives of citizens. She also distrusted the Keynesian economic model that had been in the ascendant since the 1930s. The point of this is to show that ideology plays at least as important a role in housing policy as practical issues. The size of the state, what role it should have, how much it should tax, spend and save are all ideological factors. People disagree on the role of the state, not because they particularly want their fellow citizens to live in poor housing, or not have adequate healthcare, but because they disagree about the policies that must be pursued to deliver these things on ideological grounds. Ideology is the bottom; it regards our core beliefs, and we can go no further down.

This chapter looks at the key ideologies that have shaped the past two centuries and their impact (or not) upon housing policy.

Liberalism

Liberalism is a political and economic philosophy that values human rights, equality before the law, consent of the governed, free trade and free markets. Liberals also support equal civil rights, on grounds of race and gender for example, and freedom of speech, the press and religion. Liberalism as a philosophy and practical movement gathered pace with the 'age of enlightenment' – a period during the 17th and 18th centuries where reason and science rose to prominence.

Liberalism survives in modern times at least nominally in the Liberal Democrat Party in England, and many other countries have similarly named parties that seek to maintain the liberal traditions outlined above. However, even on the so-called political left and right (embodied in the Labour and Conservative parties in Britain), liberalism is the accepted societal and political paradigm that binds. Almost no serious politician expecting to be elected would claim to be against equal civil and political rights, or against science and enlightenment.

Liberalism sought to replace the norms of hereditary privilege, an established church, absolute monarchy, divine right of kings, and traditional conservatism, replacing them with representative democracy and the rule of law (more of which in the following chapter). The English civil war was fought over the clash between the monarch and parliament, and the right to rule. It was as much a war of ideas and words as of battle and destruction (Adams, 1993). The basis upon which society should be run was at the foundation of the conflict. The medieval view was that people could be judged by their position in society and that a natural hierarchy existed with the monarch exercising the 'divine right of Kings' to rule (Aquinas, 2014; Filmer, 2016; also see the following chapter). The Tory party supported this view of the world and the right of the monarch to rule (still evident in today's Conservative Party – often still colloquialised to 'Tories'). The emerging liberal movement rejected it. Instead, they proposed creating a society run along the principles set out above. This view was represented politically by the 'Whigs', or liberal party in Parliament and more widely. In short, liberalism promotes individual freedom coupled with constitutional government and civil and human rights. The right of Parliament to rule was asserted by the Civil War but had to be consolidated a generation later in the Glorious Revolution of 1688 (Jones, 1988; Adams, 1993).

The ideas of many liberal thinkers and writers (and others) are discussed in greater detail in future chapters, but their ideas are outlined here as they are key to understanding the liberal viewpoint.

Parliamentary rule/constitutional government

Until the emergence of the liberal movement and world view, government in the UK had been via royal prerogative. There had been some early skirmishes, the signing of Magna Carta in 1215, for example, that limited the power of the monarch over the feudal barons, but the monarch ruled using the 'divine right of kings' to do so. The monarch was viewed as the 'head' of the state (which was viewed in organic terms) and governed via a network of aristocratic representatives across the country. The monarch had the final word on all matters and their decision was constitutionally binding – there remains vestiges of this in the current constitutional arrangements in the UK.

The liberal world view rejects the divine right of the monarch to rule, insisting instead that all men are created equal by their (still divine) creator, and had rights that must not be infringed upon e.g. the right to life, property and free speech. Liberal thinkers (especially Locke, 2016) were also concerned with the legitimacy of government; they rejected a divine right to rule and instead conceived of a method of government that drew legitimacy from the consent of the governed. Further, liberal thinkers recognised the dangers inherent in one person or body holding too much power, and that not only should power be diffused across the system, but the different parts of the system must also be able to hold the others to account to prevent abuses. Different powers and accountabilities have therefore been diffused across the systems of government – often known as the 'separation of powers' (this is discussed in detail in the following chapter). Suffice it to say, that in the UK, powers are separated between the legislature, executive and judiciary. The British constitution is somewhat unique in that it is unwritten, many of the rules are conventions and kept via tradition. It is unclear what would happen were they to be broken. For example, the sovereign (king or queen) signs all legislation into law (known as Royal Assent). As they hold a hereditary role and the House of Commons (HOC) is elected, the Commons has primacy and therefore the monarch has no constitutional right to refuse to give Royal Assent. But nevertheless, could do so, provoking a constitutional crisis. It's unclear what would happen if this were ever to occur. The UK system of government then is bound by a raft of unwritten rules and conventions – but still has a constitutional government whose powers are bound.

Rule of law and equality before the law

As described above and in detail in the following chapter, it is the role of government to make and pass the laws that we all must live by, within the constitutional structure. The government derives its legitimacy to do so from the consent of the people, expressed via general elections and the social contract (a tacit agreement amongst people to give up some rights in return for protection and order).

The 'rule of law' is the conceptual and legal framework that underpins an open, fair and peaceful society. It insists that public authorities be bound by, and accountable before pre-existing, clear and known, laws. In other words, the government can't make it up as it goes along. Citizens are treated equally before the law and human rights are protected. Most rights are 'negative' in that they place obligations on us only not to infringe the rights of others. There are also a variety of 'positive rights' enshrined in a number of national and international treaties that require the state to provide certain goods and services or prevent discrimination or ill treatment in relation to several protected characteristics.

Equality before the law is the idea that all men are created equal, and that no person shall hold power or control over another unless vested with it under these arrangements. For example, as already stated, we accept the authority of government and the legitimacy of the laws passed by Parliament because of voluntary elections. Economically we accept the existence of private property and the operation of markets because the exchanges that lead to these arrangements are voluntary – this is discussed in detail in its own chapter later in the book. We also accept that people have the right to make their own decisions, and live their lives as they see fit provided, they abide by the laws passed by Parliament and respect the rights of others.

Freedom of speech/the press/religion/assembly

This is perhaps the most important of all our rights, there is a good reason that freedom of speech is the subject of the first amendment to the US constitution (US, 1776); protected by the second amendment (to bear arms). The right to free speech is fundamental to our social arrangements. For example, we accept the right of people to make the case for ideas that we may well find abhorrent not because we like or agree with them, but because we accept that the best way to refute such ideas is through more speech rather than proscribing ideas or words, we find reprehensible. That said, there are some limits to free speech. It is generally accepted in both the UK and US that the right to free speech does not cover impugning the reputation of others (defamation or slander), or inciting violence against individuals or groups. In the case of the latter, it is often held that the threat of violence must be imminent and real, rather than abstract. This is perhaps best exemplified by John Stuart Mill's (2006)[1] example of the corn trader. Mill's conception of free speech accepts that any speech is acceptable except where it can cause harm. In the case of the corn dealer, Mill suggests that it is generally, perfectly acceptable to say that corn dealers starve the poor, but not so if you are in front of a corn dealers house in front of an angry mob.

It is easy to make the case that ideas or speech such as denial of the holocaust in World War II should be made illegal, indeed in Germany, denying the holocaust is against the law. However, once we accept that some speech can be made illegal (except in the circumstances discussed), we must grapple with the question of who decides what speech is allowed and what speech is not, and that the decision will probably end up being made by those people you least want to be in such a position. As an example, let's take the case of the suffragettes. Most people would agree that the suffragette's cause (equal voting rights for women) is morally righteous, and we now find it slightly surprising that anyone ever thought otherwise. However, had the case for free speech not been made in the previous centuries, who is to say that the government and other interested parties could not have written off the suffragette's campaign as beyond the pale and banned it outright?

As it was, the speech had to be allowed and they were eventually successful in their campaigning. The question is not so much, is the speech morally acceptable – the contrast between holocaust denial and the suffragettes could hardly be starker, but if we accept limitations on speech, who decides? And, what if they decided that holocaust denial is allowed (it is), but campaigning for an equal right to vote was not?

The right to freedom of speech or expression is often mentioned in concert with the freedom of religion, assembly and a free press. Freedom of religion asserts that all of us are free to follow whatever religious belief we choose to, and we should not be discriminated against because of our religious beliefs. It is also taken to mean 'freedom from religion', in that we are also free to have no religion at all, be that atheism or agnosticism, and not suffer discrimination because of it. Closely allied to this is the freedom of assembly, people have the right to assemble for religious worship and a variety of other reasons. Indeed, during the COVID-19 pandemic, cases have been brought before US state courts challenging the right of state governments to impose lockdowns and limit the right to assemble. In several cases, courts have ruled that states do not have the power to limit the right of assembly, even in a pandemic and as ill-advised as it might be to gather. For example, the US Supreme Court ruled that a cap on numbers attending religious worship (in New York), passed by the state legislature, was unconstitutional and therefore must fall.

Freedom of the press is clearly allied to freedom of speech and expression. As with these, freedom of the press mandates that government shall not take any action that limits the right of the press to print what it likes. There are some limits to the freedom of the press, but these are generally in the form of slander or libel laws that enable private individuals to challenge false statements about them in the courts, and force a correction, retraction and award of damages or compensation. Although, access to such mechanisms can be ruinously expensive and inaccessible to people of average means. The issue with freedom of the press is not that we do not accept that the press (and the wider media) has the ability and power to cause serious damage by printing false information, it is that we also accept that the result of the state acting to limit what the press can and cannot say is infinitely more dangerous; indeed, it is one of the hallmarks of totalitarian regimes throughout history.

Freedom to own property

This is the most pertinent right to the issue of housing. I do not intend to dwell on it for long here because there is a full chapter on property rights later in the book. However, suffice it to say that property rights begin with one's individual body. The right to property is not only the recognition of the right to own 'things', but the recognition that one owns one's own body, and our bodily integrity must not be violated without our consent. This recognition is the reason that slavery, assault, rape, and other offenses against the person are deemed serious criminal offenses.

Second, we have the right to own possessions, this includes a home, and we also accept that we cannot be deprived of possessions where we have justly acquired them, except in some limited circumstances and after due process. Other people cannot take our possessions without consent (theft), break into our house, garage, shed, etc. or remove anything from those places. These are all criminal offenses. There are some exceptions to these rules, that are discussed partially in the following chapter, and in the later chapter dedicated to property rights.

Reason

Key to the enlightenment and the topics discusses above is the concept of 'reason'. Enlightenment thinkers (Hume, 2004;[2] Mill, 2006; Locke, 2016)[3] rejected the assertion of religious explanations for things and instead argued that man could discover things about the world via reason. Even Thomas Aquinas in *Summa Theologica* (2014)[4] had asserted that we could make discoveries about God's plan via a mix of revelation and reason. The enlightenment rejected the revelation part, or that there was a divine plan. Indeed, even Aristotle thought that the purpose of life was to flourish or live well and discover what this means via 'right reason'.

'Reason', in this context, goes hand in glove with science. At the same time as the enlightenment, and key to it, was the development of empirical observation and positivist science. This is the idea that we can discover truths about world by experimenting and observing phenomena and conduct replicable experiments to confirm or refute hypotheses. In terms of a simple experiment, every time we have heated pure water at sea-level, it boils at 100 degrees Celsius. We therefore treat this as fact/truth. We also know from experimentation that this is dependent on constant atmospheric pressure, were this to change or become variable, the boiling point of water would change. Indeed, the boiling point of water is significantly lower at the summit of Mount Everest than it is at sea level.

'Reason' as a concept is important to our rights and responsibilities under the law, it has nothing to do with the boiling point of water, of course; but our actions under the law are often judged by what is 'reasonable'. And the acceptance that we can reason is obviously key. Indeed, an inability to reason is often used as a defence in criminal cases. Reason applies in two contexts here. First, the operation of the justice system relies on the assertion that we can use reason to discover the truth – a courtroom is effectively a crucible where we burn away irrelevancies to find the truth. Second, reasonableness is often used as a yardstick for whether our actions are lawful or not. In English law it is not unusual to hear the phrase 'the man on the Clapham omnibus', he is often used as a conceptual device to measure reasonableness.

Perhaps the most obvious use of the 'reasonableness' test is in cases of self-defence. We are all entitled to use 'reasonable force' in defence of ourselves or others from harm. What is considered 'reasonable' is dependent upon the situation, and our perception of the situation, or the seriousness of the harm that might be caused were we not to act in defence.

For example, were someone to break into your home and threaten you with a knife, if you genuinely believed yourself to be at risk of serious harm or death, it would be considered reasonable to respond with lethal force to neutralise the threat. If, however, you struck the assailant and knocked him unconscious, it would not be considered reasonable to continue the assault once the immediate threat had been removed. Ergo, a single lethal blow to an active assailant would probably be considered reasonable self-defence, killing them via a continued assault once they were unconscious would not. We now turn to conservatism.

Conservatism

Conservatism can be a tricky beast to get a hold of as it is often not seen as a coherent ideology in and of itself but rather a collection of principles and positions that revere the preservation of what is proven to work, with change only being made at a slow pace where old orthodoxies are no longer suitable or workable. That being the case, it might be useful to look at some of the prominent conservative thinkers for their views on what conservatism is. To be clear, the conservatism discussed here has little to do with the policies of the current Conservative party, the section on neo-liberalism (below) may be more suited to that.

Conservatism springs from the idea that central to our lives is the need to protect what is familiar to us (Oakeshott, 1991). Indeed, Michael Oakeshott's (1991) conception of conservatism, does not necessarily believe in a fixed set of ideas or doctrines, but rather seeks to preserve what is familiar and close. This is an argument that can be traced back to the political philosophy of Edmund Burke (See Hampstead-Monk, 2009) and is an important strand in conservative political thought. However, it is a contested view, by both opponents of conservatism and from more assertive conservatives who wish to create – or recreate – a particular form of society based on conservative principles. The emphasis on the familiar can be criticised as a mere attempt to protect vested interests, but also as a relativistic and even nihilistic position that bases itself on circumstances rather than principles. Indeed, a frequent criticism levelled by incumbent Conservative Prime Minister Boris Johnson's detractors is that he believes in nothing. Rather, perhaps what he believes in is leaving people to live their lives as they see fit, caring for that and those close to them without some grand utopian plan for society.

In his essay, 'On Being Conservative', Oakeshott (1991) details his view of the 'conservative disposition'. Oakeshott (1991) concedes that the disposition he espouses is difficult to make a strong case for. It is even more difficult to make a case for it in a political environment that venerates action and change, and when the two main opposition parties (Labour and the Liberal Democrats) both style themselves progressive. Rather, Oakeshott's conservative disposition is one of quiet satisfaction with what is. It does not seek nor motivate change, but revels in the enjoyment of things for their own sake; in that sense, it is negative. Following the footsteps of Immanuel Kant's (2012)[5] categorical imperative, those people and things we hold dear, are to be enjoyed as ends in themselves and not as a means to some other purpose.

To be of conservative disposition is to,

> prefer the familiar to the unknown, the tried to the untried, fact to mystery, the actual to the possible, the limited to the unabounded, the convenient to the perfect, present laughter to utopian bliss.
>
> *(Oakeshott, 1991, p. 408)*

To make this point, Oakeshott contrasts two types of relationships we may have with others. First, those people we call friends and those people with whom we transact business. For Oakeshott, the first of these is to be preferred. Fealty or loyalty to those we hold close trump more profitable arrangements, and we know this to be so. Should, for some reason, we lose a cherished friend or relative, we feel this loss keenly, but were our family GP or butcher to retire or close up shop, it would probably prove and annoyance and may event cause some momentary grief, but we would surely find another that would prove as satisfactory. The loss of the friend or relative would cause a far more mortal blow, there are simply some people who are irreplaceable, and we feel their absence deeply – our connection to them probably has no purpose, no profit other than for its own sake. And we feel those losses even more keenly for it.

What has been lost need not even be intrinsically better than its alternative, nor its replacement incapable of being enjoyed, but the loss is felt because that which is gone was actually enjoyed. The inclination to enjoy what is present breeds attachment and affection, the replacement offers no attachment. But that is not to say that the person of conservative disposition is indifferent to change. Indeed, their temperament demands that they be attentive to potential change and the effects of it.

Oakeshott also has this to say of change. Innovation entails some loss and some possible gain – the onus is therefore on the innovator to show that the change will, on the whole, be beneficial. The more closely a change is rooted in the situation that necessitated it, and not merely imposed on it, the less likely it is to cause feelings of loss. An innovation that is proposed as a solution to an actual problem is preferable to one that springs only from some utopian idea about the improvement of the human condition. Change should be slow and incremental, and attention paid to the results of that change with a view to making course corrections en route; and finally, any change is best timed when it will be limited to what is intended, and not be corrupted by unintended consequences or mission creep.

The proper role of government, therefore, is to protect the people who wish to be left alone to enjoy their lives, from those who seek to alter their way of living. For Oakeshott, a government is an Aristotelian endeavour that seeks to leave each of us to discover our own version of the 'good life' or, as John Gray (2009) has it, our *modus vivendi*; and to then leave us to enjoy it. The government is there to enforce the rules of the game on all alike and not force one man to live by another's rules. Oakeshott asks us, are those who rule endowed with a superior wisdom which discloses to them a better range of beliefs and activities with which to impose on everyone else? What makes one man's dreams superior to another's? We can tolerate monomaniacs, but why should we be ruled by them? The government must therefore accept the diversity of opinion and activity and not impose upon subjects one of their own. The role of the government is not the management of enterprise, but rather the management of a great diversity of enterprises (Oakeshott, 1991). Change and modification of the rules should be considered and incremental. Such modifications of the rules should always reflect and never impose change on those that are subject to them.

Civil society, limited government and the nation state

In the above section on liberalism, I have already covered the role of the constitutional government and the rule of law in maintaining a modern democracy. I have also mentioned the role of conventions in the UK's unwritten constitution (and there is further discussion in Chapter 2 also). For conservative thinkers e.g. Michael Oakeshott (1991), John Gray (2009), and Roger Scruton (2001; 2006) (and many others), civil society is just as key to the governance of the nation, any nation. By civil society, we mean the mass of custom, tradition, common law, canon law and other social institutions and arrangements that enables or limits the conduct of citizens without recourse to statute or the law courts. Indeed Gray (2009) posits that Thomas Hobbes (in Leviathan) was correct and that without a government and a social contract, the state of nature would allow the "unalterable defects of the human condition" (p.135) to flourish and life would indeed be "nasty brutish and short" (Hobbes, 2014).

However, unlike Hobbes who believed that an all-powerful government (the Leviathan) was acceptable to end the state of nature, central to the conservative philosophy of Oakeshott (1991), Gray (2009) and Scruton (2001; 2006) is the limited role of the government. This is an acknowledgement and acceptance of the imperfection of the human condition. Central to conservativism is a belief in a limited form of government that accepts this imperfection (Gray, 2009).

> Politics is a way of coping with the hazards that come from the unalterable defects of the human animal.
>
> *(Gray, 2009, p. 135)*

However, Oakeshott (1991) is clear that one may be of conservative disposition in respect of the government but quite radical in respect of other matters. Gray (2009) sees Oakeshott's conservatism as a bulwark against the utopianism of both left and right. For Gray, the proper role of the government is,

- The avoidance of civil strife
- Containment of war
- Mitigation of arbitrary power via the rule of law
- Provision of a sound currency.

Gray (2009) rejects any utopian blueprint (of left or right). Conservatives should be committed to the reform of institutions and practices that do not serve human interests, and the aim is to do this via increments rather than grand schemes. The costs of changes must be assessed to ensure it moves with the grain of national character and tradition rather than against it. The role of tradition in modifying and controlling behaviour is to enable people to make choices that comply with learned wisdom (Gray, 2009). Gray fears a situation developing where all matters are left to legal remedy, asking what of a society that lives with difference and doesn't resort to the authority of the courts to resolve differences, this leads to a situation where one party 'wins' and the other 'loses', one point of view is ruled 'correct' and the other in error.

The third strand of the gordian knot is the nation state. A functioning civil society and limited government can only exist and maintain legitimacy via a stable nation state. Scruton (2006) defends the nation state not on the grounds that it is the only option to resolve problems, but that it is the only one that has proved itself. The nation state, patriotism and limited government create the conditions of loyalty and shared national endeavour that enables a population to bear the weight of the expectations of civil society and the law.

Those of conservative disposition (according to Oakeshott, [1991] at least) seek to modify, and thus improve, the nation state by increments rather than monumental change and to leave the people to discover and fulfil their own version of the *modus vivendi*. As Kekes (1999) says, conservatives are motivated to preserve the conditions that enable them to enjoy their own lives, and to extend these conditions to all others. For the more people there are with little to lose, the more they will threaten the present condition, upon which present and future good depends.

Communism/Socialism

Like conservatism, socialism suffers from something of a definitional problem. Many who claim espouse socialism seem to believe that it merely entails the higher taxation of those considered to be 'rich', to fund services for both those considered poor, and for the population in general. In short a 'progressive' taxation and spending system. This view is not entirely without merit, in the post–World War II Britain, much of the political discussion has been about how much to tax the rich and what should be available for the poor. However, post 1979 and the election victory of Margaret Thatcher, the Conservative Party (1951) has arguably adopted a much more ideological approach to public spending and services, abandoning the so-called post-war consensus.

Karl Marx (with Friedrich Engels) (2008)[6] is generally thought of as the originator of socialism and communism. In his studies of conventional philosophy, Marx rejected pure philosophical enquiry. He saw the task of trying to understand the way things are as falling short, instead believing that the point of studying social conditions should be to change them, rather than merely understanding (Graham, 1988). Further, Marx found pure philosophical enquiry 'idealistic' in that it concerned itself with coming up with good ideas, ignoring social conditions 'on the ground'. He therefore proposed that what was needed was a rigorous and theoretically conceived idea of the forces of historical processes. Marx thought that there were two such processes, the means for people to provide for themselves (survival) and the growth in population. To meet the need for survival and to enable the growing population to also survive, people form what might be called 'systems of production' to exploit the resources around them, for mutual benefit (Graham, 1988). This continues until the 'systems' created can no longer produce enough to the meet the basic needs of all, and then the system must be changed for one that is able to do so. Therefore, there will be two groups of people. One whose needs are adequately met by the current system (who naturally wish to maintain it) and one group whose needs are not met and therefore favour the current systems demise and replacement with a new one that can meet the needs of all. Marx saw that this antagonism was the engine that drove history down its teleological course. This is often said to be the fundamental contradiction within capitalism that would eventually lead to its downfall and replacement.

Marx found that as a system of production, capitalism could produce infinite goods, thus ending one of the driving forces of social change. However, he also found that rather than being able to support infinite needs (driven by population growth), capitalism actually meets the needs of fewer and fewer people. Under a capitalist system, wealth is distributed highly unequally between the owners of the means of production and the workers. As the population grows, more and more people are drawn into the 'working' class and provide the ownership class with greater and greater reward (Graham, 1998).

The result of this is that most people will be just as unable to meet their basic needs in a system of infinite production, as in previous systems that could produce enough; thus, setting up a confrontation between the ownership class, who wish to maintain the current system, and the workers, who wish to overthrow it (Graham, 1988). Marx therefore thought the overthrow of capitalism inevitable, and what would emerge would be an end to the historical forces that had determined history to this point, and replacement of this with a 'classless society'. In this classless society (Communism), all needs will be able to be met, the problem of private ownership and unequal distribution of resources will be solved; therefore no one will have a vested interest in overthrowing the system.

Karl Marx's writings (together with Engels) are significant and dense, this is therefore a somewhat simplified version of his theories. But, enough to understand the origins of modern socialism and how it is represented in modern Britain via the Labour Party, Trade Unions and other left-wing organisations.

Modern socialism in the UK (sometimes known as British socialism) is embodied in the Labour Party. Labour campaigns and manifestos generally aim at higher taxation on the 'rich', increased redistribution of wealth via the welfare system, increased spending on public services, and increased state ownership of some industries; most recently a proposal to renationalise the railway network in the Labour Party's 2019 election manifesto (Labour Party, 2019). During the post-war consensus years (c.1945–1979), much of British industry was state controlled, for example British Rail (trains), British Leyland (motor manufacture) and British Steel. Whether British socialists are/were ideologically committed to bringing about full communism via State ownership or not, state ownership of those industries thought to be 'public goods' is a key part of their existence. It is worth saying here that the UK Labour Party is a relatively new political party, formed in the mid-to-late 19th century. It was formed to give representation to the urban working class that had gained suffrage via the *Representation of the People Act 1884*. This was also wrapped up with the formation of the Trades Union Congress (TUC) at about the same time. Until this point, political opposition until this point had been mainly between the Whigs (liberals) and Tory party (the forerunner of the modern Conservative Party).

The British Labour Party has a long tradition of radicalism and campaigning for the rights of working people. The Labour movement should be seen as an Aristotelian enterprise, based upon the 'good life', natural law and human rights – the search for the common good by common endeavour (Glasman, 2011). The Labour movement drove significant advances in the rights of workers during the 19th and 20th centuries. Advances in worker conditions, health and safety, paid time off, pensions etc. were all achievements of the Labour movement during this time. Some municipal services e.g. the burial movements, building societies and co-operatives pre-dated the welfare state, in the case of the NHS, collective funding for healthcare provided Aneurin Bevan with the blueprint. The post–World War II vision was of a society that is cohesive, generating a common good and a strong sense of looking after the welfare of all. The strongest bearing the brunt of the cost, whilst caring for those unable to protect themselves (see Beveridge Report, 1942).

Following World War II, pressure built for a new social deal. With roots in the Marxist and Fabian socialist traditions (Foord & Simic, 2005), the post-war consensus (as it became known) and related housing policies can broadly be said to be in the 'social democratic' tradition. The post-war 'social democracy' had equality at its heart (Crosland, 1956). It sought to harness the economic growth produced in a mixed (public and private) economy with social policies designed to produce greater equality; both of opportunity and outcome. In doing this, social democracy provides a mid-ground between democratic and radical socialism (Jay, 1986) acknowledging that the homogenous, capital owning class predicted by Marx had not come about.

To implement this, more equal, vision of society based upon incremental growth and Keynesian economic management (see Keynes, 1936) the government took control of the means of production for those things most closely related to the provision of infrastructure. In terms of housing, health, welfare and education, citizens were no longer subject to market forces and held a stake in society via the non-capitalist (nationalised) industries and systems that recognised 'need' as a variable that should play a role in the allocation of resources rather than allocation by market forces. In the period after World War II, the Atlee government embarked upon the setup a huge bureaucracy designed to provide cradle to grave social support to all (see Hill, 1996; Marquand & Seldon, 1996; Baggott, 2004; LaValette & Pratt, 2006; Dean, 2008), a grand 'social democracy' with massive investment in council housing, the NHS, and nationalisation of key industries.

Critics of the Beveridgean welfare settlement of the post-war years found that the welfare state replaced the sense of mutual responsibility, evident in the burial movements etc. with rights to universal benefits. Key figures in the Labour movement (e.g. Atlee, Bevan and Morrison) had sprung from municipal socialism (Brendan O'Neill Show, 2019) but then moved away from 'worker run' industries in favour of state management. This meant a preference for state planning over 'worker management', Glasman (2011) complains that the managers in the nationalised industries were *PPE graduates*, not workers and nationalisation of key industries placed management prerogative as the organising principle, the unions therefore became antagonistic . For Glasman (2011), the democratic associations that Labour champions should exist as an independent bulwark against the worst aspects of free market dominance – an alternative source of power to promote the common good and the Labour movement exist as a counterweight to capital, a partner to it and not subservient.

The Labour Party is now in danger of losing hegemony over the class it was created to serve (Rutherford, 2011). Rutherford bemoans Labour's dismissal of working-class culture. In an echo of John Gray's *modus vivendi*, for Rutherford, culture is a narrative and tells the story of who we are, Labour should be the party of the common good and the common life (Rutherford, 2011). Indeed, he sees the role of the Labour movement as to mount a militant defence of the common life, as well as individual liberty and of an ethical life; the Labour movement must work against both the commodification of life by the market and usurpation of common life by the state. The left-wing struggle for liberty was always a struggle for democracy, not paternalism (Rutherford, 2011).

In a UK context then, socialism has departed from pure Marxian theory in that it has pursued socialism via state ownership of (some) of the means of production, provision of council housing, healthcare, social care etc. via state-run services. One of the key breaks in the Marxist tradition has been between those favouring revolution, and those favouring an evolution into socialism. It has been argued that state provision of services, such as in the UK, has held back progress towards socialism by providing comfort for the working class within a capitalist paradigm thus preventing them forcing the move to Marx's classless society (Ravetz, 2001). To the chagrin of some (Glasman, 2011; Rutherford, 2011), state provision of services and state-based rights to welfare benefits and services have replaced working people providing these things for themselves via mutual and co-operative societies and municipal provision.

Other ideologies

'Liberalism', 'conservatism' and 'socialism' have been by far the most influential ideologies in terms of UK politics; hence they are covered here in detail. There are, however, other ideologies that are of lesser influence and therefore, import; but are worth briefly covering here both for the sake of completeness, and for the fact that they have had considerable influence.

Neo-liberalism

This is perhaps the most influential school of thought in the modern era, and as it sounds is a return to liberal principles, mostly in the realm of the economic. Although the term has been around since the early part of the 20th century, in a UK context, it has largely gained relevance since the 1970s and the reforms to public spending made by the Thatcher governments of the 1980s. For some critics, this continued through the Blair and Brown governments of the 1900s and 2000s, and into the coalition and Conservative governments since 2010. In its simplest form, neo-liberalism can be said to be the resurgence of laissez-faire capitalism. It was a shift away from the Keynesian economic consensus of the post-war era and marked by tax reductions, deregulation of capital markets, privatisation of services and reductions in both government spending and involvement in people's lives.

Neo-liberals favoured the theories of monetarist economists such Milton Friedman (2020) and Friedrich Hayek (1990; 1997). These thinkers believe that the public (and public servants) are motivated to act by their own self-interest, and that 'public interest' or its wider ideology is a fallacy; the only true democracy being that of the market. Therefore, the best way for people to fulfil their needs and wants is for them to make free choices in the marketplace (see Irving-Clarke, 2019) – monopolistic services provided by government are clearly antithetical to this ambition. Where services are provided by government, the use of quasi-market mechanisms to procure and manage public services can be used to ensure that public sector managers are focused on the needs of 'consumers' rather than themselves.

This relies on two drivers, first the consumer acting selfishly to fulfil their own needs/desires and exercising choice between services thus driving the 'market'. The second driver is created by the government itself imposing requirements and targets upon officials thus driving them to act in their own self-interest in fulfilling these targets. As Seddon (2008) puts it,

> Public sector managers have a surrogate profit motive, a target that defines success, and the incentive to earn investment in their organisation and recognition among their peers.
>
> *(p. 7)*

Savage and Atkinson (2001) draw a direct line between these theories and the emergence of "New Public Management" (NPM) in the UK from 1979 to the present. (The term New Public Management was coined by Christopher Pollitt in 1990.)

NPM pays greater attention to results or outputs and the personal responsibility of individual managers (Hughes, 1998); it emphasises a move from tall hierarchies to flatter, more flexible and devolved management structures with more flexible (deregulated) terms of employment (Horton & Farnham, 1999). Institutional and individual objectives are clearly defined and include the establishment of key performance indicators (KPIs) against which to measure delivered outputs (from Hughes, 1998; Horton and Farnham 1999). There is greater attention to economy, efficiency and effectiveness as greater commitment to procedures such as market testing and compulsory competitive tendering (from Elcock, 1991; Hughes, 1998). Finally, there is a general commitment to reducing the role of the state in the disposal of gross national product and the provision of services in favour of privatised companies and the establishment of a market in areas previously viewed as the preserve of natural monopolies or public administrators (from Hughes, 1998).

The third way

The New Labour governments of Tony Blair (then Gordon Brown) (1997–2010) were first elected in 1997 when the country seemed exhausted of the Conservatives, and the Conservatives seemed exhausted (Dale, 2021). Blair swept into power with a huge majority in Parliament and a mandate for his reforming agenda. Blair talked of a 'third way', between the first (social democracy) and second (neoliberalism) ways. In many ways, Blair followed a neo-liberal agenda, retaining a focus upon the global economy, reform of the welfare state, a drive for modernisation and a focus upon targets, indicators and audit and inspection regimes (Newman, 2001; Barber, 2008; Seddon, 2008). However, there was also greater complexity in that the (New) Labour governments "attempted to establish – and sustain – a new set of political alliances. It sought to forge a consensus around an agenda of 'modernising' reforms designed to remedy deep-seated social problems" (Newman, 2001, p. 1). Newman also finds a softening of the ideological commitment to market mechanisms and competition as drivers of public sector reform. There was emphasis upon innovation, experimentation and policy evaluation in public services backed by a commitment to evidence-based policy in many areas (Nutley et al., 2007).

Labour's modernisation agenda also went beyond the mainly economic approach of neoliberalism. There was a reimagining of the cultural and social spheres. New Labour saw their mission as having to reconcile a retention of the economic gains of the Thatcher governments with a reshaping of civil society via the invocation of moral and civic values, and a new emphasis on citizenship and social inclusion. They sought to appeal to new groups, women, black and minority ethnic communities, disabled people, LGBTQ+ groups and the young; groups that they saw as excluded from the political system. At the same time, they sought to play down the post-war (second way) class-based enmity – which they saw as a corollary of "Old Labour" (Newman, 2001).

The third way can be defined against the two prevailing ideologies of the 20th century, but also as embodying both previous systems, promoting social justice, the market and community. It could embrace market enterprise but not automatically favour market solutions, promote a positive role for the state but not assume the state should provide services directly; and offer a communitarian view of society rather than an individualistic one. Blair's position was that the third way represented a modern social democracy, a path on the centre left of politics and an opportunity to advance centre left values whilst utilising policies that were suited to the modern world. Blair's view was therefore that the third way was not a path between left and right (a middle way), between the first and second ways; but a new strand on the left (Blair, 1998).

The third way is light on ideology and heavy on pragmatism (unlike its first and second-way predecessors). It was values that mattered to the third way project rather than individual policies or how they are delivered. Blair (1998) outlines these values as equal worth, opportunity for all, responsibility (correlated with rights) and community. The third way is not tied to the state (first way) or the market (second way) as mechanisms of delivery, but rather seeks to work with the voluntary sector, the government (local and national) and other agencies so solve the complex social problems that government faces (Driver & Martell, 1999).

Conclusion

At the start of the chapter, to I asked the simple question (to paraphrase) 'Why can't we agree that the government should provide housing for all?'

I have started this book by looking at ideology, as it is the basis for what we believe. There is nothing below it. Ideology can be seen as the assumptions we make about the world that form the basis for our view of it. It is also connected with the way we see human nature (ontology), what constitutes knowledge (epistemology) and moral imperatives (axiology). It is particularly often seen as the basis upon which form economic or political theories and the policies to implement ideas.

In this chapter, I have outlined what I consider to be the main political and economic ideologies in practice today. I have also tried to delve into the background of each of these to identify the underpinning ideas and logics.

To try to answer the question asked, I hope I have shown that there are some distinct ideological differences between people and groups that directly affect their views on the question. Those on the political 'left' would probably be more than happy with the government providing for all. Their vision of a world with greater equality fits in well with everyone living in housing provided by the state regardless of their ability to pay for it. Indeed, Aneurin Bevan's express vision was one where people from all walks of life, from doctors and lawyers to manual workers would live together in housing provided by the state. This is evident in the post-war Labour government's distinct policy choice to embark on the large-scale building of homes using local authorities to do so and socialising large amounts of land that was previously owned privately (King, 1996). It is also true to say that those on the 'left' are more comfortable with high government spending and a generally more interventionist state.

If one were being cynical, it would be easy to suggest that the lack of security of tenure in post-war council housing was a mechanism to maintain power of tenants. This was only introduced in council housing in the *Housing Act 1980* as the Conservatives sought to level the playing field between council and housing association tenants, and private renters. Indeed, we would rightly baulk at the idea that anyone should be made to leave their home without due process – in this case a court hearing. In a strange piece of symmetry, the current government has promised to end so-called no fault evictions. These are evictions where the initial protected period of an assured shorthold tenancy has ended, and the landlord can ask for the tenancy to be ended without giving a reason. Further, this is what is known as a 'mandatory ground', this means that when the case comes before a judge, they have no choice but to grant an order returning the property to the landlord; meaning the tenant can be evicted. Campaigners have often cited this process as not sufficiently fair or robust, given the serious consequences for the tenant who may not have done anything wrong. Property rights are discussed more fully in Chapter 5.

This brings us to what is the main difference between liberals (as described) and small and big 'c' conservatives – the political right. The 'right' are first more sceptical about state spending than the 'left', and more sceptical about state intervention generally. More specifically, Conservatives have a distinct view of housing and home that sets them apart from those on the left. For the 'right', housing is a bulwark against the state. The saying 'An Englishman's home is his castle' has passed into the realms of cliché, but the point it makes is still a good one. Our housing, or homes, are first the main economic asset that most of us own. As I discuss in Chapter 5, owners have a 'deed' to prove ownership of their land and the house that is on it. This deed is virtually unchallengeable, in the following chapters on the state and property rights, I look in detail at this. Somewhat ironically, although home ownership is a bulwark against the state, it also requires the state to recognise and enforce not only the right to own property but also the legitimacy of 'deed' documents that prove ownership. Indeed, in the chapter on property rights I look at a comparative example of countries where this is not the case. One of the main sources of business start-up capital is borrowing against the home, making it one of the main sources of independence and wealth creation also.

Aside from economics, housing also provides a bulwark against the state in that it provides a place of privacy where we can perform our basic freedoms (King, 2003). King makes the case because we are not truly free unless we have somewhere private to do this, housing can be seen as a negative right, rather than a mere socio-economic claim against the state and our fellow citizens. Further, if we think about what it means to have freedom, one of things that is essential is the separation of the public and private self. What this means in practice is that we are able to have a private self to which the state is not privy. That we have a private place where we are able to express our own thoughts and feelings, where the state is unable to encroach; and even where they are, are powerless to take action against us. This idea is perhaps best expressed in the First Amendment of the U.S. constitution, when it says that the government shall make no law that abridges the right to freedom of speech or expression.

If we think about books and films that depict totalitarian states, *Animal Farm* (Orwell, 1945), *1984* (Orwell, 1949), and *V for Vendetta* (2005), they depict societies where there is no division between the public and private self. In 1984, *Big Brother* has devices in every home so that it can spy on what they are saying and hold people accountable for 'Thoughtcrime'. This is enforced via a Bentham-esque or Foucauldian (Foucault, 1995) 'panopticon' where one never knows when one is being observed, and so modifies one's behaviour to never break the rules for fear of discovery and punishment. There is no private self.

This is not merely the realm of books and films. The totalitarian regimes of the 20th century also obliterated the private self. In Nazi Germany, the Gestapo secret police sought out wrong thinkers, the Stasi in East Germany did similar. One only has to read Alexsandr Solzhenitsyn's epic account of the USSR under the Soviet Party (Solzhenitsyn, 2018) to see how the Soviet Cheka obliterated the private self. Neighbours spied on neighbours and were rewarded for turning in wrong thinkers and those even mildly critical of the regime. In the gulag too, inmates were rewarded for informing on their fellow prisoners, until it was their turn. Where the private self no longer exists, totalitarianism does.

The home then acts as a bulwark against the state. It is where we are allowed to be our most private selves, to express our private thoughts and feelings to one another away from the rest of the world. It is where we are allowed to be our most authentic selves. The psychiatrist Carl Jung put forward the idea of the *persona* (Jung, 1947). The *persona* or mask is the outward face we present to the world, thus disguising our real selves. It is our public face, different to who we really are. Our home then is where we can let the mask slip and be out most authentic selves.

This has not gone unnoticed in parts of the left. Engels (1884) identified the home and the family as key sites of 'oppression' and the recreation of capitalist hegemonic structures. It has therefore long been the target of Marxists and the left more generally. One only has to look at communist architecture or the brutalist concrete blocks that appeared across many western countries in the 1950s and 1960s. Brutalism was a deliberate attack on aesthetics and the home. Le Corbusier and his ideas about home as a machine was the same. 'Soviet Man' has long been a construct of the socialists and communists. Leon Trotsky (1924) wrote about the 'communist man', this basically meant an adherent of Marxism-Leninism, a man not ruled by his base instincts but by conscious self-mastery and hard work. The Neo-Marxists also took this idea on, writing that man must transform himself into a new 'social man', we are not even able to conceive what the new Marxist society will look like, so corrupted is the present society. Brutalism was a crude attempt to effect this change, the aesthetic of the old world had to go in favour of uniform, concrete carbuncles. For the (neo) Marxists, equity is the ideal end state, everyone must live in the same type of home, with the same state provided appliances – see Trabant cars in the old East Germany for example.

What I hope I have shown here is that there is not any disagreement that housing, or home, is a positively good thing. Rather, there are different views about the nature of housing and what it is for. For those on the left, housing is a resource that is perfectly well provided by the state, and in the interests of wealth distribution and social justice, the state should provide good quality and affordable housing to people unable to access the market, and for those further on the left as a means of achieving equity and the prerequisites for what they believe will deliver the utopia detailed by Karl Marx, Friedrich Engels and others.

For those on the right, they do agree that good quality, affordable housing is a good thing. However, they are also moderately to highly sceptical of the state as a medium to provide it. For conservatives, the home stands for something more than merely somewhere to live. For them, it is a bulwark against the state, a place of privacy where we can be our most authentic selves. I am not suggesting that those on the left don't think this also, but it is more a feature of conservative policy.

Economically also, those on the left are generally more comfortable with state spending, and government being a landlord. Whereas on the right, people are generally against excessive state spending and think that the government interference in the free-market should be minimised.

Notes

1 Original date of publication is 1859.
2 The original date of publication is 1751.
3 The original date of publication is 1689.
4 The original publication date is unknown.
5 The original date of publication was 1785.
6 The original date of publication is 1848.

Suggested further reading

Blond, P (2010) *Red Tory: How Left and Right Have Broken Britain and How We Can Fix It*, Faber, London.
Crosland, A (1956) *The Future of Socialism*, Jonathan Cape Ltd., London.
Oakeshott, M (1991) *Rationalism in Politics and Other Essays*, Liberty Press, Indianapolis IN.
Scruton, R (2006) *A Political Philosophy*, Continuum, London.

References

Adams, I (1993) *Political Ideology Today*, Manchester University Press, Manchester.
Aquinas, T (2014) *Summa Theologica*, Catholic Way Publishing [Online].
Baggott, R (2004) *Health and Health Care in Britain*, Palgrave Macmillan, London.
Barber, M (2008) *Instruction to Deliver: Fighting to Transform Britain's Public Services*, Methuen Publishing, London.
Beveridge Report (1942) *Social Insurance and Allied Services*, UK Government, London.
Blair, T (1998) *The Third Way: New Politics for the New Century*, Fabian Society, London.
Blond, P (2010) *Red Tory: How Left and Right Have Broken Britain and How We Can Fix it*, Faber, London.
Brendan O'Neill Show (2019) *Maurice Glasman: Blue Labour and Brexit*, https://audioboom.com/posts/7531949-maurice-glasman-blue-labour-and-brexit.
Conservative Party (1951) *Conservative Party General Election Manifesto*, Conservative Party, London.
Crosland, A (1956) *The Future of Socialism*, Jonathan Cape Ltd., London.
Dale, I (ed) (1991) *The Prime Ministers*, Hodder & Stoughton, London.
Dean, H (2008) *Social Policy*, Polity Press, Cambridge.
Driver, S & Martell, L (1999) Left, Right and the Third Way, *Policy and Politics* 28(2) pp.147–161.
Eichengreen, B (2007) *The European Economy Since 1945: Coordinated Capitalism and Beyond*, Princeton University Press, Princeton NJ.
Elcock, H (1991) *Change and Decay: Public Administration in the 1990s*, Longman, Harlow.
Engels, F (1884) *The Origin of the Family, Private Property and the State*, Zurich.
Filmer, R (2016) *Patriarcha*, Createspace, Scott's Valley CA.
Foord, M & Simic, P (2005) *Housing, Community Care and Supported Housing – Resolving Contradictions*, CIH, Coventry.
Foucault, M (1995) *Discipline and Punish: The Birth of the Prison*, Random House, New York.
Friedman, M (2020) *Capitalism and Freedom*, University of Chicago Press, Chicago IL.Glasman, M (2011) Labour as a radical tradition in M Glasman, et al. (eds), *The Labour Tradition and the Politics of Paradox*, The Oxford London Seminars, London.
Graham, G (1998) *Contemporary Social Philosophy*, Blackwell, Oxford.
Gray, J (2009) *Gray's Anatomy: Selected Writings*, Allen Lane, London.
Hampstead-Monk, I (ed) (2009) *Edmund Burke*, Ashgate, Farnham.
Hayek, F.A. (1990) *The Constitution of Liberty*, Routledge, Abingdon.
Hayek, F.A. (1997) *The Road to Serfdom*, Routledge, Abingdon.
Hill, M (2006) *Social Policy in the Modern World: A Comparative Text*, Wiley Blackwell, Hoboken NJ.
Hobbes, T (2014) *Leviathan*, Wordsworth Editions, Stansted.
Horton, S & Farnham, D (1999) *Public Management in Britain*, Palgrave, Basingstoke.
Hughes, O.E. (1998) *Public Management and Administration*, MacMillan, London.

Hume, D (2004) *An Enquiry Concerning the Principles of Morals*, Kessinger Publishing Company, Whitefish MT.

Jay, R (1986) Democracy in R Eccleshall, V Geoghan, R Jay & R. Wilford (eds), *Political Ideologies: An Introduction*, Hutchinson, London.

Jones, J.R (1988) *The Revolution of 1688 in England*, Weidenfield & Nicolson, London.

Jung, C (1947) *On the Nature of the Psyche*, Ark Paperbacks, London.

Kant, I (2012) *Groundwork for the Metaphysics of Morals*, Cambridge University Press, Cambridge.

Kekes, J (1999) *A Case for Conservatism*, Cornell University Press, London.

Keynes, J.M. (1936) *The General Theory of Employment, Interest and Money*, Palgrave Macmillan, London.

King, P (1996) *The Limits of Housing Policy: A Philosophical Investigation*, Middlesex University Press, London.

King, P (2003) *A Social Philosophy of Housing*, Ashgate, Farnham.

King, P (2016) *The Principles of Housing*, Routledge, Abingdon.

Labour Party (2019) *It's Time for Real Change: The Labour Party Manifesto 2019*, The Labour Party, London.

LaVellette, M & Pratt, A (1996) *Social Policy: Theories, Concepts and Issues*, SAGE Publications, Thousand Oaks CA.

Locke, J (2016) *Two Treatise of Government*, Createspace, Scott's Valley CA.

Marquand, D & Seldon, A (1996) *The Ideas that Shaped Post-war Britain*, Fontana Press, London.

Marx, K & Engels, F (2008) *The Communist Manifesto*, Pluto Press, London.

Mill, J.S. (2006) *On Liberty and the Subjection of Women*, Penguin Classics, London.

Newman, J (2001) *New Labour, Policy and Society*, SAGE, London.

Nutley, S.M., Walter, I, & Davies, H.T.O. (2007) *Using Evidence: How Research Can Inform Public Services*, Policy Press, Bristol.

Oakeshott, M (1991) *Rationalism in Politics and Other Essays*, Liberty Press, Indianapolis IN.

Orwell, G (1945) *Animal Farm*, Secker and Warburg, London.

Orwell, G (1949) *1984*, Secker and Warburg, London.

Ravetz, A (2001) *Council Housing and Culture*, Routledge, Abingdon.

Rutherford, J (2011) The future is conservative in M Glasman, et al. (eds), *The Labour Tradition and the Politics of Paradox*, The Oxford London Seminars 2010/11, London.

Savage, S.P. & Atkinson, R (2001) *Public Policy Under Blair*, Palgrave, London.

Scruton, R (2001) *The Meaning of Conservatism*, Palgrave, Basingstoke.

Scruton, R (2006) *A Political Philosophy*, Continuum, London.

Seddon, J (2008) *Systems Thinking in the Public Sector: The Failure of the Reform Regime and a Manifesto for a Better Way*, Triarchy Press, Bridport.

Solzhenitsyn, A (2018) *The Gulag Archipelago (Abridged edition)*, Vintage Classics, London.

Trotsky, L (1924) *Problems of Life*, Methuen, London.

V for Vendetta (2005) [Film], James McTeigue dir, USA, Warner Brothers.

2

WHY DO WE NEED THE STATE ANYWAY?

Case study

Following a period of housing crisis, a new government is elected in the UK. The new government is elected on a manifesto promise to provide good quality, genuinely affordable housing for all, no matter what it takes.

The new government uses its mandate and parliamentary majority in the HOC to take unprecedented measures.

- The planning system is completely abolished, the government gives itself the power to plan and build what it thinks is needed, wherever it likes without the need for local consultation or any other measures.
- The government gives itself the power to compel owners to sell second homes to local people at a government-determined price.
- The government gives itself the power to compel all households to move to a home that suits their household size and needs – according to a government needs assessment – including people with accessibility needs moving to adapted homes.

When challenged by opposition MPs and in the courts, the new government argues that people have a right to be well-housed. There is no written constitution and therefore no formal limits on its powers, the measures were in its election manifesto, and it has a majority to override objections and amendments in the Lords, it can and will do what it likes.

DOI: 10.1201/9781003298212-3

Questions

- Does resolving the housing crisis and ensuring that everyone has a secure, affordable, and suitable home justify these measures?
- Is the government right that it can do as it pleases, with no limit?
- What rights is the government breaching by acting in this way?

What does the state do, why do we need it?

The modern state touches nearly every aspect of our lives. The state comprises the government and opposition members of Parliament (MPs), the House of Lords (HoL) and the Monarch in Parliament. Beyond that, the state employs an army of civil servants that work in government departments, the police, security services and health service. Local government also employs many people providing local services, museums, roads, maintenance and of course, council housing. By extension, the government and local authorities also have a significant degree of control over other aspects of life through funding and regulation. The degree to which this exists is one of the main faults that runs through politics and society more widely. As I set out in Chapter 1, there are various political ideas and ideologies that have been popular at one time or another. Some remain popular and form the basis of the ideological differences that we see in politics and society more widely. The role of the state is key to these differences. But where did the state come from? Why does it exist? And how do we decide what its role is?

In theoretical terms, what one perceives the role of the state to be can largely depend on the assumptions one makes about what came before its existence. In philosophical terms, this is often referred to as 'the state of nature' (Hobbes, 2017;[1] Locke, 2016).[2] Thomas Hobbes and John Locke are perhaps the two best known 'state of nature' theorists, but they have very different conceptions of the state of nature, and this means that their view of the state is also very different. Preceding Hobbes and Locke, and the source of their concerns was the 'divine right of Kings' to rule. The discussion therefore begins with Thomas Aquinas' view of the state and its function(s) before moving on to Hobbes, Locke and others.

Aquinas and the state

Feudal society (in the UK at least) was governed by the combination of church and state (which are still inextricably linked in the UK by the established church). Aquinas (in Summa Theologica [2014])[3] sets out his view that society is constructed by God, and man is bound by the design that God has laid out. Unfortunately, Aquinas also believed that God's design was beyond the understanding of men, but that through reason and the bible, we could discover part of that design (Fink, 1981). For example, when a rock falls to earth, it obeys God's law for inanimate objects. Man, on the other hand, is viewed (correctly) as a rational being, capable of independent thought. The more rational man became, the closer to God he moved. Aquinas also knew that man could act independently and in accordance with his own will. Acts against God's law were therefore punishable in accordance with the same law. For example, theft was clearly against God's law as revealed in the ten commandments and therefore rendered the thief liable for punishment. The role of the state was therefore to discover and interpret God's (natural) law and punish or reward people accordingly (this is still reflected in Canon Law today).

The analogy of the human body was used to construct Aquinas' social theory. The family was likened to the body under the head(ship) of a father, the family was to obey the instructions of the head – a hand or foot cannot question instructions from the brain. However, the head was also not to forget that it is attached to the rest of the body and make decisions for the benefit of the whole. This analogy was continued with 'peasants' being the children of the Lord of the Manor with the same 'bodily' rules in application. Everyone in the country were viewed as the children of the monarch – the ultimate power in the country. However, the monarch was not all powerful and could not simply act as they wished. They too were limited by natural law and God's reason, such as it could be understood.

Hobbes and the leviathan

Hobbes took a somewhat different view. Writing in the context of the carnage of the English Civil War, in Leviathan, Hobbes sets out his view of the state of nature and of the role of the state in regulating society. Whereas Aquinas believed that without social institutions, people would still exist within a pre-ordained and hierarchical framework, Hobbes thought that they would exist only in a 'state of war'. The state of nature for Hobbes was one where there were no pre-ordained rules or structures. Life was "nasty, brutish and short" (Hobbes, 2017). There was no reason why one person could or would not assault or kill another, there was no good cause to cultivate land or acquire property, as one could never be sure it would not be appropriated by others, by brute force. However, he also recognised that the strongest person could be easily killed in their sleep, even by a child.

For Hobbes then, the purpose of social institutions is to protect everyone, from everyone else and prevent life being a complete free for all. To do this, Hobbes proposed that a person should give up their right to harm others, on the condition that everyone else do the same. This is known as a 'social contract'. Further, this meant that a sovereign entity must exist to punish individuals who break the social contract. Ideally (for Hobbes), this should be an individual monarch with absolute power (the titular Leviathan). The sovereign retained the sole right to use violence on others and was not subject to the social contract. The sovereign retained all executive, legislative and judicial power, no other person had any power to question or punish the sovereign. However, Hobbes also provided that where the sovereign could not retain the power to protect subjects from harm, the people could voluntarily leave that sovereign and appoint another who could do so. So, for Hobbes, the power of the sovereign is total, he/she is a monarch with absolute power and control over individuals and society more broadly. In the context of UK society today, we would clearly find this incompatible with ideas about freedom and personal autonomy. Yet, why is the idea of an all-powerful sovereign so bad if it keeps everyone safe and ensures that society runs smoothly and in the best interests of all?

Locke and the social contract

John Locke (2016) agreed with both Aquinas and Hobbes in some respects. He agrees with Aquinas that there is a creating God and that all people are created in an ordered society where they have certain rights and duties to one another. He agrees with Hobbes, that there exists a 'state of nature' and that the state has an important role in ensuring that the behaviour of people and groups do not infringe upon the rights of others. However, he also disagrees with Aquinas and Hobbes in important ways.

Locke wrote two treatises of the government that set out his political philosophy and these have been hugely influential on governments in the UK and US, and other places too. In his first treatise, Locke addressed Roger Filmer's *Patriarcha* (Filmer, 2016),[4] in which Filmer agrees with Aquinas' conception of the social order, that society is constructed according to a natural order that assigns status to everyone in the society with the monarch at the head, with lords (and ladies) beneath them, and finally 'serfs' at the bottom of the social order. For both Filmer and Aquinas, this social order is pre-ordained, and the role of the monarch is to maintain this according to God's law. The church is, of course, key to this view of society and exists to divine, interpret and enforce God's natural law.

Whilst Locke agrees that there is a creating God and a social order, he disagrees with Filmer and Aquinas that this is a hierarchical social order. Indeed, his first treatise if government is widely considered to be a destruction of Filmer's' *Patriarcha*, and the idea of the 'divine right of kings' was not really taken seriously in the UK again following the Lockean demolition.

In his second treatise of government, Locke sets out his political philosophy and view of the social contract. He begins by defining political power as

> the right of making Laws with Penalties of Death, and consequently all less Penalties, for the Regulating and Preserving of Property, and of employing the force of the Community, in the Execution of such Laws and in defence of the Common-wealth from Foreign Injury, and all this only for the Publick Good.

Much of his second treatise flows from this definition. For Locke, there is a distinct moral dimension to how society should be ordered and governed. Unlike Hobbes, Aquinas and Filmer, Locke is not content with an arbitrary or pre-ordained total authority of a monarch to rule and is instead concerned with a moral justification for the wielding of power. Locke agrees with Filmer and Aquinas that there is a social order and that there is a state of nature, however Locke disagrees that there is a hierarchy, arguing instead that all are created equal and subject only to God's law, also known as 'natural law'. Further, all are responsible for enforcing this natural law – and it is this duty that confers the right to punish offenders. Locke sets out three natural laws. First, the right to life – everyone is entitled to live, second, the right to liberty – everyone is entitled to do as they wish, is so far as it does not conflict with the first right; and lastly, the right to estate, the right to own all one creates, gains through gift or trade so long as it does not conflict with the first two rights. These are summarised within the constitution of the US as the "right to life, liberty and the pursuit of happiness" (US, 1776).

Locke states that these natural laws are universal in application and contraventions must be punished. However, unlike Filmer and Hobbes, Locke is of the view that exercise of arbitrary power would lead to injustice and violence. He therefore proposes that this can be avoided if everyone agrees to consent to recognise a civil government with the power to enforce the laws of nature – this is Locke's version of the 'social contract'. For Locke, a primary function of society is to set up a framework within which legitimate contracts, those which do not violate the laws of nature, can be enforced.

Rousseau, the general will and the state

Jean-Jacques Rousseau (1998)[5] provides a third and distinct account of the state and its purpose. Rousseau, like Locke and Hobbes, agrees that there is a social contract and that this contract is in place to ensure that all persons are free from harm and having their rights violated by others. However, Rousseau believed that the freedoms we enjoy as individuals are essential to our humanity and to give up these freedoms would compromise this. He also believed that there are significant gains to be made from a social contract – banding together to agree to a sovereign state that can protect us and uphold our rights. So, Rousseau is entertaining two opposite ideas, first that our inherent freedom is a good thing and intrinsic to our being, and second that we should give up some of this freedom, and by implication some of our humanity for a greater good. Rousseau sets out to explain how we can both retain our freedom and participate in a state. Key to this his idea of the 'general will'.

The general will is the will of the state as a whole and aims at the common good (Warburton, 2010). It is perhaps best understood in opposition to the 'will of all'. It may be the will of all the people combined for a certain action to happen, for example, it may be the 'will of all' to cut taxes as they will all individually benefit from having more money due to the cut in the tax rate; however, it may be best for society as a whole to maintain the current rate as it provides the state with greater resources. The second of these is taken to be the 'general will' and what should happen. Rousseau, therefore, draws a sharp line between selfish, individual interests and the interest of society. There is no scope to derogate from the general will, it is for the good of society to comply with it and depends on us setting aside our individual freedoms and desires. At first reading, this can seem like a strange definition of freedom and Rousseau does indeed say that even where we may disagree with the general will, we must be 'forced to be free', or comply with the general will.

One of the main problems with Rousseau's 'general will' is how we discover what it actually is. If it is not the same as the aggregate of the 'will of all', how can we possibly divine it? In the UK, we have had some experience of attempts to find out recently with referenda that have been held on the voting system Scottish independence and membership of the European Union (EU). To take the latter, we have taken the result of the referendum to leave the EU as the general will (in Rousseau's terms) and enacted it. But some 'remainers' have also argued (again in Rousseau's terms, if not their own) that remaining is actually in the best interests of the country and therefore the 'general will' – the referendum result should therefore be discarded as merely the 'will of all'. Although it could equally be said that regardless of the economic and social impacts of leaving, the democratic vote to do so must be respected and it is this that is actually in the best interests of society – the impact of the government ignoring the referendum result which it had promised to enact would be severe. So, which of these is the 'general will'?

Robert Nozick and the 'nightwatchman' state

Thus far we have looked at three theories of the social contract that state that a society must band together for protection against harm, although their conceptions of how this works are very different. But, although these theories speak of the formation of a 'social contract' as a historical event, they also all treat it as a theoretical one. There is no moment in which a social contract is formed, but rather its existence is a conceptual device to help us to understand our current social and political arrangements.

So how does the state come to exist and why? Why is it self-evident that such an event is inevitable? The libertarian theorist Robert Nozick in *Anarchy, State and Utopia* (1974) gives an account of how this may occur, drawing also on social contract theory. It is worth rehearsing Nozick's argument here as it takes the formation of the state out of the theoretical realm and provides a practical, if also theoretical, example of the formation of a state apparatus.

As a starting point, Nozick agrees that the 'state of nature' is a dangerous place and there are no reasons why all of us are not at risk of having our property stolen, being turned off our property or being assaulted or killed. His solution to this is therefore that individuals will band together to provide protection of their properties and persons. Much like NATO, collective agreements form to protect life, liberty and property with a group of individuals promising to come to each other's aid should their rights be infringed. Further, some individuals and households will, instead of participating in such agreements, wish to purchase the service from others. Nozick theorises that this will lead to the formation of numerous 'protection services' that act as *de jure* police forces. Further, he thinks that as these services merge and some become significantly larger than others on account of their effectiveness, eventually there will be few large protection providers. Eventually, either one of these will out, or it will become desirable for these to combine into one organisation providing protection to everyone in the social group. At this point, Nozick says, the society will have, to all intents and purposes, formed a state. That said, in *Anarchy, State and Utopia*, Nozick argues for a minimal state that acts to protect the rights of the individual and enforce contracts, which are key to his theory of justice (see Chapter 3). He also argues, as does Locke, that we should only act where we can do so without infringing on the rights of others and uses the concept of 'side constraints' to show how this is possible and desirable. It should be noted that whilst Nozick provides a practical and compelling account of how the state comes into existence, and is indeed inevitable, he only ever intends this to be a minimal or nightwatchman state, existing only to safeguard rights. He does not intend it to become as complex as modern states have become and get involved in most aspects of life.

The problem with social contract theory

The existence of a social contract is persuasive and the basis of the social and legal systems in the UK, US and many other countries. The social contract and indeed the rule of law is predicated on the idea that people are sovereign individuals with equal worth before the law, and in most cases should only be held accountable for their own actions and not those of others, nor should the state or anyone else indulge in collective punishment for individual crimes.

However, from where does the social contract originate? If we think about the terms that we usually expect from a contract, e.g. that it is freely entered into, legally binding, fair and made in good faith, how many these apply to the social contract?

Is the social contract freely entered into? Well, no person has the power to dictate when or where they are born, indeed birth is complete accident of fate. If one is born in the UK, US or somewhere else, one then finds oneself born into the social contract that applies in that country or state. From this perspective, one does not freely enter into a social contract but ends up in one as an accident of birth. It is also not easy to leave either. One could in theory move to another country where the social conditions are more appealing and indeed this has happened in the past. Many thousands of people left the UK to travel to the Americas in search of a new life. However, they moved to a nascent country of largely uninhabited land where a new life was there to be forged (more of which in a later chapter). In the modern world, there are very few lands left to settle and leaving one country for another is not easy as most have restrictive immigration policies. It seems unlikely then, that the social contract can be said to a voluntary arrangement.

Second, is the social contract legally binding? Well, it certainly is on the citizen. One must obey legitimately passed laws including payment of taxes and not infringe upon the rights of others.

One of the key features of a contract is that, once signed, it is legally binding on both sides. Can this be said of the social contract? The state, or more accurately the government, has the legitimate power (by virtue of democracy) to repeal laws it finds out of date, or otherwise uncongenial, and make new laws via the legislative process. For example, between 1945 and 1979, it could legitimately have been said that the social contract included the provision of housing by the state for those that wanted it, as well as a raft of other social democratic measures including the ownership and management of key industries. However, in the years that followed, there was wholesale change to this as the Thatcher governments privatised industries and transformed the social housing landscape by a raft of financial and social policies that residualised social housing to housing only those that could not compete in the market. The state has the power to change the terms of the social contract whenever it wishes to, and this is not consistent with what we would generally think of as a fair agreement. This is perhaps an argument for limiting the power of the state as Nozick (1974) suggests. The fewer things the state controls, the fewer things it can change.

Limiting the power of the state

Above, I have offered several theoretical models of the state, ranging from Hobbes' leviathan state to Nozick's minimal 'nightwatchman' state. The fact of the matter is the UK lies somewhere in between these extremes. Of the most use to us at this point, is the thinking of John Locke (2016). Locke is rightly concerned with the moral justification for a coercive government and how the powers of that government are constrained so that it does not strain beyond its proper function of protecting the rights he enumerates. At this point it is worth considering what we mean by rights by briefly discussing positive and negative rights (see Berlin, 1969).

Negative rights

These are the rights we usually think of as such. For example:

- The right to life
- The right to private property
- Freedom of speech and expression
- Practice of religion
- Self-determination.

These are rights that only require others to respect them and take no action that infringes upon them. For example, all that it takes for me to respect your right to life is for me not to kill you; or the right to freedom of speech only requires that others do nothing to infringe upon it.

Positive rights

The second category of rights are called positive rights or sometimes legal or citizenship rights. A positive right is a right that you have to a good or service that requires action from another person or entity to provide it. For example, the right to housing or health requires that some entity acts to provide those goods or services. Housing requires that some person or entity plans, constructs and pays for it. Likewise, healthcare is a service provided by an individual or organisation and therefore people need to train and work as doctors, nurses etc. to ensure that this right (if it is a right) can be fulfilled. Perhaps one of the biggest debates between the political left and right is the extent to which positive rights are rights at all. As the current discussion is regarding the limitation of state power, it is more the former type of rights that we are concerned with, although the limitation of the state can impact upon its ability to deliver on the latter.

In the UK, as it many other countries, the power of the state is limited by a constitution. More unusually, much of the British constitution is unwritten and is a collection of rules, conventions and other mechanisms that have built up over time. This is both a strength and a weakness. Contrast this with the US that has a written constitution as the founding document of the country. Whilst the UK constitution is not codified, the Supreme Court recognises several key constitutional principles, including parliamentary sovereignty, rule of law, democracy and upholding international law. These are worth brief discussion here.

Parliamentary sovereignty

Exactly as it sounds, this is the idea that Parliament is sovereign – that is to say it is the highest source of law in the land, via Acts of Parliament. This sovereignty was hard won through battles between the crown, courts, church and the people. The Magna Carta (1215) (see Carpenter, 2015) asserted the right of a 'common counsel' (now Parliament) to exist and limited the 'divine right of Kings' (Aquinas, 2014; Filmer, 2016) to levy taxes on baronial lands. The Magna Carta also guaranteed common land to the people for farming, fishing etc. In 1534, King Henry VIII asserted his divine right over the catholic church in Rome and cemented the monarch as the head of the Church of England, thus ending the formal rule of the papacy in Great Britain. In 1615, the Earl of Oxford's case asserted the authority of the Court of Chancery over the common law courts, meaning that the courts could no longer strike down Acts of Parliament. Finally, the glorious revolution of 1688 and establishment of the Bill of Rights in 1689 asserted Parliament's sovereignty over the monarch. These arrangements persist to the present day.

Rule of law

As with the constitutional arrangement, there is no statutory definition of the rule of law. However, as per Lord Bingham of Cornhill, it is generally taken to mean that the law will be clear and predictable, not subject to broad or unreasonable discretion, applies equally to all people, has speedy and fair procedures for enforcement, protects fundamental rights and upholds international law. As ever, some of these terms are not defined, and so the application of the rule of law continues to develop via case law.

Democracy

This is the bedrock of both the rule of law and parliamentary sovereignty. The implementation of a functioning representative and deliberative democracy is what gives these other principles legitimacy (as per Locke), and is the opposite of arbitrary rule. Arbitrary power is exercised by one person alone (as per the divine right of kings), whilst in a democracy, the power to install or remove those that rule them lies with the people.

Institutions

To safeguard the rights of the people, the state has been divided into three institutions: the legislature, executive and judiciary. Each of these institutions has a set of powers that are limited in both scope and to that institution. Although this kind of arrangement was mentioned by Aristotle in Politics (2009),[6] where he drew upon the constitutional forms of the time, the Roman republic divided into Senate, Consuls and Assemblies, its modern form is commonly ascribed to Montesquieu. In *The Spirit of Laws* (1748) (see Montesquieu, 2015), he describes the separation of political powers into the three now familiar institutions. These are illustrated briefly illustrated below with some comparison to the US system. As Margaret Thatcher famously observed, Europe was created by history, America by philosophy.

The executive

In Montesquieu's original formulation, the executive power rested with the monarch. This is reflected to this day by the necessity for the monarch (in the UK) and the president (in the US) to sign all acts that passes the legislative process – known as Royal Assent. However, in modern times, this (in the UK) has been only a convention (an unwritten rule that shall not be broken). Legally, there is no reason that the monarch should not refuse to give royal assent, but to do so would provoke a constitutional crisis. The modern arrangement is that the power of the executive sits with the cabinet in Parliament, along with Crown prerogative to take certain actions – going to war has been one such action for example, although this has been challenged in recent years.

The executive has the power to propose legislation and take it through the legislative process in Parliament. It also has the power to implement those laws that reach the statute book and to implement policy via the executive functions of the government. However, as astute readers will have noticed, the Cabinet in the UK is made up of MPs, who are also part of the legislature. To this degree, there is not a strict separation of powers between these two bodies. Contrast this with the US where the executive power sits with the president, who (as in the UK) signs all legislation into law, but also retains a (partial) veto. The president also oversees the executive agencies of government and is responsible for implementing laws and policy.

The legislature

This is part of the state that is generally elected by the people and responsible for formulating and passing legislation (laws) into statute. When a law is proposed by the executive (or occasionally via a Private Member's Bill), members of the legislature have an opportunity to amend or defeat legislation by voting on whether it should pass into law or be changed before it does so.

In the UK, and some other countries, the legislature is divided into three parts. These have different names in different places, but generally there is a lower house, an upper house and the sovereign (head of state) (see above). In the UK, the lower house is the HOC, the upper house, the HoL, and the head of state is the reigning monarch. The UK has some very specific constitutional arrangements relating to these parts, which are worth considering briefly.

House of Commons

In the UK, the HoC is the only part of Parliament that is elected by the people of the country. For that reason, despite sometimes being called a 'lower house', it is generally considered to be the primary of the three parts. This is because it draws its legitimacy from the consent of the governed via general elections as per Locke's (2016) thoughts on the legitimacy of government.

House of Lords

The HoL was historically made up of peers of the realm appointed originally by the monarch but more recently by Parliament. All political parties have an opportunity to recommend appointments to the Lords. Historically, peerages were hereditary and so places in the HoL were inherited by the eldest living (male) child of a peer when they died; and indeed, there remain to this day several Lords who continue to sit in the HoL on this basis. There are also a number of 'life peers' who sit in the HoL and are able to vote on legislation. Life peers are generally former politicians or people with knowledge of expertise on a given area that are placed in the Lords on that basis. Of course, there are frequent and justified criticisms that peerages are a source of patronage for people who do political favours. As suggested above, the Lords is seen as secondary to the HoC. Indeed, after a running dispute, the HoC asserted its primacy via the *Parliament Act 1911*, which enable the HoC to override the HoL in some circumstances.

The sovereign

The final constituent part of Parliament is the sovereign or, in the UK, the monarch. The monarch holds very little actual power in the UK as it has been severely circumscribed by convention and the development of democratic government. There are several Crown prerogative powers that are generally held by the Executive in Parliament and can be exercised by such. The role of the monarch is now generally ceremonial and limited to requesting the leader of the winning party to form a government, dissolving parliament at the request of the Prime Minister (although this has now been curtailed by the *Fixed Term Parliament Act 2011*) and signing statute into law. The powers of the monarch are also strongly limited by convention. For example, there is no reason the monarch could not ask the leader of the losing party to form a government, dissolve parliament when he/she chooses or refuse to sign a Bill into law. However, to do so would breech convention and provoke a constitutional crisis. Again, the non-elected status of the monarch is the reason for this limitation of powers.

Judiciary

The judiciary is the institution of the state charged with interpreting the laws that Parliament passes and deciding how they should be implemented. There are different parts of the judiciary that cover criminal, civil and administrative law. The judiciary is also made up of several layers of courts to allow for an appeals system. For example, in the criminal law, there are magistrates' courts, Crown court, the Court of Appeal and Supreme Court. Judges are also appointed offices and are generally life appointments to avoid political patronage. The role of courts is also limited in that it has no control over what laws Parliament may pass or the process; although it does have a role in the 'Judicial Review' of laws and policy, ensuring that they are lawful and do not impede the rights of the individual. Until 2005, the highest court of appeal was the HoL, with several 'Law Lords' sitting in the upper house for this specific purpose. However, this role has now been separated and given to the Supreme Court – one of the main arguments for this was to maintain the separation of powers. Although, the Secretary of State for Justice is also the Lord Chancellor, the most senior legal officer in the Country. This means that head of the justice department sits as part of the executive and legislature, thus impeding the separation of powers.

A comparative note

To provide a point of comparison, in the US model (probably the closest to the UK), the model is very similar, but different in some key ways; and it is useful to understand these and it illuminates the strengths and weaknesses of both models.

In the US, the Houses of Congress (HoC) are the legislature. Congress is divided into two houses, the House of Representatives (similar to the HoC) and the Senate (roughly equivalent to the Lords). Both houses have specific powers enumerated in the constitution. They also carry out different functions. For example, the House of Representatives is currently made up of 435 directly elected members. The number of representatives for each state is calculated by their population with each 'district' having a single representative. Because of this, the House of Representatives is said to represent the interests of the people – much like the HoC in the UK.

The Senate (or upper house) is the rough equivalent of the HoL, but importantly, is directly elected but on a different basis to the House of Representatives. Each state elects two senators, there are currently 50 states, therefore 100 senators. The senate is therefore said to represent the states' interests rather than the peoples'.

Finally, the sovereign or head of state function is fulfilled by the Office of President of the US. The president heads the executive branch and appoints the executive agencies that implement the law and policy; the president is responsible for directing these agencies. The president also heads the investigatory agencies, e.g. Federal Bureau of Investigation and Central Intelligence Agency, and is commander-in-chief of the armed services. The president is elected via the electoral college, a mechanism set down by the constitution to prevent the highly populous coastal states from dominating presidential elections.

Because the two HoC are both elected, they are considered co-equal in powers (unlike the UK where the primacy of the sole elected house has been asserted in both statute and in convention). This means that are occasions where there is gridlock between the two houses where they cannot agree. This is often the case on finance bills and has led to government shutdowns and furloughing of federal staff. Contrast this to the UK, whereby convention states that the HoL does not contest finance bills, or bills relating to manifesto commitments.

As we have already said, in the UK the monarch cannot, by convention, refuse to sign into law a bill that has passed the legislative process. In the US however, the elected president can veto a bill by refusing to sign it. The president's veto can be overridden by a two-thirds vote in both houses to do so. However, where Congress has adjourned and the president vetoes a bill, there is no mechanism to overturn this.

Despite the similarities then, the two systems are also very different. One of the main strengths of the UK system is that it allows the elected government to implement the manifesto it was elected on – and this is important. However, in the US system, all three parts of the legislature have legitimacy as a result of being elected and separation of powers is much more strongly observed. It should not be forgotten that the founding of the US was via a bloody revolutionary war based upon the objections of some Americans to paying taxes to the King of England to shore up Britain's exchequer. As the founders were fighting against the exercise of arbitrary power by an unelected monarch, they were understandably concerned about arbitrary abuse of power and too much power being held by one individual or part of the government. Hence, they designed a system that not only divided power between the branches of government, but also built a level gridlock into the system so no one branch could dominate the others. Although, one of the strengths of the UK's unwritten constitution has been said to be its ability to be flexible and cope with wider changes in the world.

Justifying the state

From the theoretical and practical discussions above, the existence of a state can be justified, both theoretically and practically. The existence of an entity that can make laws, protect individuals and groups, and mediate disputes seems to be an entirely sensible proposition. The question therefore is what (limited) powers this entity should hold and how this can be justified morally and ethically.

We might start by asking, 'why is it self-evident that Hobbes was wrong?', and why shouldn't there be an all-powerful monarch who can punish people who transgress on other people's rights? Why should they be elected and not appointed, or even self-appointed? The answers to these questions lie in the assumptions we make about humanity and our intrinsic worth.

The basis of Locke's ideas about the state of nature, rights and the moral case for justifying the powers of the state stems from the idea that all people are created equal and with intrinsic worth and rights. This is most evident amongst the opening lines of the US' declaration of independence (US, 1776),

> We hold these truths to be self-evident, that all men are created equal, that they are endowed by their Creator with certain unalienable Rights, that among these are Life, Liberty and the pursuit of Happiness. That to secure these rights, Governments are instituted among Men, deriving their just powers from the consent of the governed.

This paragraph states clearly that all men are created equal, with some rights that may not be infringed upon, and that the government may be derived from the consent of the governed for the protection of those rights. The rights mentioned explicitly here, life, liberty and pursuit of happiness; and those enumerated in the bill of rights (the first ten amendments) are *negative* rights in that all that is required is for none to act in such a way that infringes upon them. The role of the state, therefore, is to prevent such infringement and to punish offenders, and obtain restitution for those infringed upon. The UK has no such founding document, and the role of the state is therefore far less defined and limited. Instead, as

I have described above, the organs of state have developed far more organically than in the US via a mix of pragmatism and philosophical thought. In both cases, however, there are a similar set of institutions designed to check and balance each other and hold one another to account.

So, why can't I just do what I like?

In the discussion above, we discussed Hobbes' conception of the 'state of nature' as a 'state of war', where people are free to do as they like. Hobbes' fear that this would lead to complete free for all where people are free to maim and kill others, take their property etc. led him to conclude that only an all-powerful sovereign could bring order, mediate disputes, and punish offenders against others. Aquinas and Locke (albeit from very different viewpoints) also conclude that people cannot just do as they please. For Aquinas, this was limited by natural law as revealed by the Church and the status hierarchy from Sovereign to serf. Locke agrees that there are some natural laws that people cannot be allowed to break. These also extend from our being created by God and that his wish is for us to live out our natural lives free from our natural rights being infringed. Likewise, very few (if any) societies allow people to do exactly as they like – this is because we recognise that everyone is of value and has rights that must be respected. The remainder of the chapter looks at some aspects of housing policy as case studies demonstrating the conceptual material we have discussed.

First, at a macro level, the state plays a key role in planning and funding housing. This is achieved via national government issuing the National Planning Policy Framework (NPPF) and provision of funding for local authorities and the construction industry to build new housing, sometimes at risk.

Planning

Planning is the process by which we decide what can be built and where. The central government sets the National Planning Policy Framework (NPPF), which is then implemented by local authority planning departments who have some discretion within the national framework. Both central and local government derive their authority to make such decisions from the fact that they are elected by the people to represent them, mediate disputes, and make decisions on their behalf.

For example, you are not allowed to build whatever you like even if you own the land you want to build on, lock stock and barrel (King, 2016). This is because building can seriously impact upon others and their right to the 'quiet enjoyment' of their property. For example, building a large extension might block the natural light to their property, mean that you can see into their home or garden and thus invade their right to privacy. Neither do we allow (in most cases) new buildings that are out of keeping with the local area. Building a skyscraper made mainly of steel and glass would likely be completely acceptable in the middle of

London (see the Shard, for example), but not in the middle of a Cotswold village where most of the properties are older housed made of local stone. A skyscraper would doubtless cause distinct distress to the denizens of the Cotswolds. The state therefore has the discretion and power to prevent this.

There is another side to this, however. Just as we do not allow people to build what they like, there are also times when the state must use its powers to override the feelings of local people in the planning process. A key part of the planning process is the opportunity for people affected by a planning application to raise objections. This can be done on a number of grounds including those already mentioned, e.g. the proposed development is out of keeping with the local area, or cutting out natural light. Other reasons local people might object to a planning application might be that it will increase the level of traffic to an impractical or unacceptable level or materially change the wider locality. Indeed, it is not unusual to see requirements for large developers to improve local infrastructure, e.g. widen roads, or build in adaptations to block noise or maintain the environment. However, there are occasions where local people use their ability to object to developments to block them, only because they do not want any building in their local area. This has become known as 'NIMBY-ism' (Not In My Back Yard).

As well as being able to make decisions that override local concerns, the state also has the legitimacy and powers to make broader decisions about the national interest. This is most often seen in large infrastructure projects that are completed for the greater good but can impinge upon the rights of individuals to a huge degree. This has been seen most recently in the HS2 project which has been the subject of controversy for exactly this reason. In conducting such projects, government has the power to use the land that it owns for such projects as well as compulsorily purchase land from private organisations and individuals that are necessary for the completion of the project. However, as this is an egregious infringement upon the property rights (see Chapter 5) of the individual, government power is also subject to the checks and balances described above. For example, large scale infrastructure projects are first subject to the parliamentary process, plans must be approved by both Houses, and become legislation; this may also give ministers powers to carry out certain actions via statutory instruments. However, these use of statutory instruments and powers such as the planning process and compulsory purchase orders are in turn subject to judicial review, to ensure that the government are acting lawfully and within the constitutional powers they have.

Regulation

Just as we are concerned that new buildings are reasonably in keeping with the local area and does not interfere with the rights of others, we should also be concerned that buildings are constructed according to agreed safety standards. The price of this not being the case has been graphically illustrated by the fire at Grenfell Tower. Buildings that are unsafe may collapse, cause injury or be otherwise unsafe.

Within the planning system there are also different requirements for different types of buildings. For example, different planning consent is needed for residential buildings than for buildings used for retail or as a pub or restaurant. The reason for this is that these uses require very different regulation. For example, a private home with three bedrooms will likely only be occupied by four or five people at the most. The residents will only be cooking for themselves, and only normal residential activities will be taking place. Contrast this with a building of similar size that is being used a retail space.

A retail building will potentially be much fuller with less free space and more potentially flammable material etc. It will also likely have more people in it at any one time, shops can get very full, and it seems that is no upper limit on how many people can enter a single shop. For these reasons, planning requirements for shops are more stringent that for homes. Clearly a home must be suitable for that use, it must be structurally sound, weatherproof and the gas and electricity supplies meet standards. In the case of electricity, there must be a system (the fuse box) that can automatically shut off the power in the event of a fault. However, a retail space may need a fire system (alarm and/or sprinklers) for the safety of customers and staff. It may also need more exits to enable many people to escape quickly in the event of an emergency.

Similarly, the building regulations contain a specific section (part M), for housing for people with disabilities. In addition to requirements for safety and comfort, we also accept that some people with, particularly physical, disabilities need a level of adaptation to be able to carry out everyday tasks that enable them to live independently. For example, housing for people that use a wheelchair must meet space standards to enable them to manoeuvre comfortably around. Kitchens and bathrooms may have lowered surfaces or adapted taps etc. Some bathrooms may even be full 'wet rooms'. The state therefore has accepted a role in ensuring these needs are met. More recently, campaigners and organisations that house people with disabilities have pushed for all (or most) housing to meet a 'lifetime standard' so that it is suitable for anyone throughout their life course (see Habinteg, 2011).

Civil liberties

A key function of the state, according to social contract theory, is the protection of civil liberties. The state acts to protect the civil liberties of the individual from others who may infringe upon them, whilst the legislature, executive and judiciary check and balance each other's powers to ensure that the state does not infringe upon them either.

However, there are occasions where the state may need to infringe upon individual rights for the protection of the individual or other people. The most obvious of these is when the state send an individual to prison for committing a serious criminal offence. This may be partly as a punishment, for rehabilitation and/or for public protection. The state can also continually imprison or detain people where a court, advised by experts, deems them to be a risk to themselves or others because of their mental health or some other reason specified in law.

In years past, the state took a rather less enlightened approach to housing those who could not provide for themselves and their families. Under the 'Poor Law', those who could not house themselves were first criminalised by the Vagrancy Act 1547, which was updated in an act of the same name in 1824 – this remains on the statute book to this day. The Act for the Relief of the Poor 1597 established a relatively simple framework for the distribution of monies, food, and other essential items via *Overseers of the Poor* (Irving-Clarke, 2019). This was updated in 1601 to enable the creation of almshouses and poorhouses (see Panell & Thomas, 1999 for a full history).

Housing arrangements for the poor then generally split into housing for those that were able to work and those that were not. For those able to work, Houses of Industry provided housing and food in return for this 'industry', in reality, this was nothing more than indentured servitude to the state. For those who refused to work to pay for their keep, Houses of Correction awaited – for the 'impotent poor'. Pauper children often became apprentices in the emerging industries (Irving-Clarke, 2019). In the 19th century, housing for the poor became bound up with the Industrial Revolution that was transforming the country and industry. At this time, broader changes were making profound change to society. The enclosure of 'common' land forced those who had previously lived an arable existence into the emerging cities to work in factories. Although an abuse of state power, it meant that many workers were better off working in these factories than the skilled artisans who had lost their livelihoods to the new industrial processes – Fink (1981) states that 500,000 weavers lost their livelihoods following the introduction of the powered loom – many perished from starvation.

The new upper class of factory owners had gained particular political power at this time (Fink, 1981), and partly used this to get those who disrupted the workforce removed. 'Lunatics' as they were termed, but in reality were often people with mental health conditions or learning difficulties, had become a significant disruption to the workforce (Murphy, 1991) and were therefore otherwise housed. The Victorian period saw an explosion of construction of asylums to house people unable to contribute to work productively. These were often externally impressive buildings, monuments to civic provide. The interiors, in contrast, were often made up of drab corridors. Murphy (1991) argues that these were designed to be therapeutic environments where a system of rewards and punishments could be used to treat patients. In reality they were often places where families could offload those they considered a burden, Scull (1993) coined the term 'warehouses of the unwanted' to describe them.

Growing numbers of people were being incarcerated in asylums, by 1809, there were over 86,000 certified cases of insanity in England and Wales, a fourfold increase over the past 45 years. Further, Malin et al. (1980) quote figures saying that 90,000 people were detained in 1939, peaking at 148,000 by 1954. Safeguards on detention were minimal (Irving-Clarke, 2019) with women being detained for having children outside wedlock and other dubious causes.

Even as housing policy developed with council/social housing growing because of the two world wars (see above), tenants lacked tenancy security and could, in theory at least, be asked to leave at any time for any reason (Cowan & Marsh, 2018), although they do note that this was not common practice. The state retained considerable power over council tenants, for example, subject subsidies (housing rebates or housing benefits paid to the individual) were not introduced until the 1970s (King, 2006), subsidy was all on the supply side (object subsidy). Until the *Housing Finance Act 1972*, local authorities did have the power to charge reduced rents, but this was purely discretionary and another lever that could be pulled to alter tenant behaviour. In this, there are distinct echoes of Octavia Hill's paternalism and use of continuing tenure to coerce tenants into behaving in a manner congenial to her own sensibilities.

Anti-social behaviour

Following the granting of security of tenure to council and other social tenants, the threat of eviction could no longer be used as a method to alter behaviour considered deviant or disruptive (notice here a commonality with the creation of asylums to remove the disruptive), the concept of 'anti-social behaviour' (ASB) began to gain traction. This isn't to say that there aren't genuine problems with ASB to be dealt with. In a seminal review of the role of social housing, Hills (2007) reported that over a fifth of social tenants reported the presence of drug users or dealers and the general levels of crime, fear of being burgled, vandalism and litter, as a serious problem. Eighteen per cent of social tenants stated they felt unsafe when alone at home or even outside in daylight. But ASB became a catch-all term for any behaviour considered problematic, or outside societal norms (Cooper, 2005).

During the 1980s, the (Thatcherite) idea that the post-war housing model was not working coincided with rhetoric on anti-social behaviour and people living in social housing gaining traction. The Conservatives introduced ASB into housing policy via the *Housing Act 1988*. This gave local housing authorities the power to obtain injunctions against perpetrators of anti-social behaviour, with a power of arrest attached to injunctions where there was actual or threatened violence. Social landlords could apply to court to demote an assured tenancy on the grounds of ASB where there was

> Conduct that is capable of causing nuisance or annoyance to some person (who need not be a particular identified person) and that directly or indirectly relates to or affects the landlord's housing management functions.
>
> *(Housing Act, 1988, c.68)*

This step cemented anti-social behaviour as an issue that impacted on housing management functions and has resulted in a continued focus by landlords ever since. To further regulate anti-social behaviour, the *Housing Act 1996* introduced a new regulatory framework for Registered Social Landlords (RSLs) and gave local authorities discretion to use introductory tenancies for all new tenancies which made it easier for them to evict tenants who exhibited anti-social behaviour within the first year of their tenancy. The focus on anti-social behaviour continued into and accelerated under New Labour Governments (1997–2010). The introduction of the *Crime and Disorder Act 1998* (redefined ASB as behaviour that causes "harassment, alarm or distress to one or more persons not of the same household as the person". This act also introduced the Anti-social Behaviour Order (ASBO) which became synonymous as defining the state of modern Britain (Squires, 2008).

The *Anti-social Behaviour Act 2003* (Section 12) amended the *Housing Act 1996* to place a duty on social landlords to publish their anti-social behaviour policies and procedures (this came into force in 2004), so that tenants and members of the public could have information about the measures that landlords would use to address ASB issues. ASB was accepted as part of the language with housing and increasingly became part of the apparatus to address it; housing workers were seen as visible authority figures alongside the Police.

In the General Election of 2010, a coalition government was elected in the UK with the Conservatives as the majority partner and the Liberal Democrats in the minority. The coalition government had a markedly different approach to ASB than their Labour predecessors (Heap, 2016). The coalition under the prime ministership of David Cameron with Teresa May as home secretary pursued a strategy based upon families and individuals as the primary victims of ASB, Labour had concentrated far more on communities as a whole (*ibid*).

There was a twin-pronged approach with Cameron championing families via the 'Troubled Families Programme', whilst May emphasised the individual and repeat victims of ASB. Heap (2016) notes that this was in sharp contrast to their predecessors Blair and Blunkett who worked closely together to define Labour's approach (Garrett, 2007). The 2014 Act replaced the Anti-Social Behaviour Injunction with the 'Part One Injunction' that could be used against a person aged ten or over if two conditions were met, first that the court was satisfied, on the balance of probabilities, that the respondent had engaged or threatened to engage in ASB and second, that the court considered it just and convenient to grant the injunction for the purpose of preventing the respondent from engaging in ASB. ASB is defined in the act as,

- Conduct that has caused, or is likely to cause, harassment, alarm or distress to any person;
- Conduct capable of causing nuisance or annoyance to a person in relation to that person's occupation of residential premises, or;
- Conduct capable of causing housing-related nuisance or annoyance to any person.

A Part One Injunction can require the respondent to do anything described in the injunction (positive requirements). It can also include the power to exclude persons from the home in cases of violence or risk of harm regardless of tenure. A range of agencies can apply for an injunction including a local authority and a housing provider and a chief officer of police for their area.

Such an injunction allowed landlords to work closely with their tenants by imposing positive requirements as injunction terms, which could enable landlords to take effective action to address behaviour. Positive requirements may include attending a substance misuse treatment organisation for assessment and support and could also include a requirement to attend a domestic abuse perpetrator programme for assessment of suitability for the programme. A local authority or housing provider could apply for an injunction against a perpetrator that contained a positive requirement to engage in a programme to address their behaviour. Breaching the terms of such an injunction gives rise to a mandatory ground for possession of the property. However, a court would expect that informal approaches had been initiated before formal, legal action was taken; although the legislation recognises that where professionals get to the point that formal action is necessary, they should have the ability to take that action quickly. Such an injunction relies only on the civil burden of proof (balance of probabilities).

Since the institution of the rule of law, in taking punitive action it has been normative to expect a law or legal agreement (e.g. a tenancy agreement) to define the scope of an offense and then for an impartial arbiter(s), such as a magistrate or jury, to judge a person's actions against this objective standard. In the discussion above, I have tried to show how this system has emerged over a considerable amount of time. As a society, we have moved away from a structure that gives one person at the top of the hierarchy absolute power over others lower in the structure. The landlord-tenant relationship is usually governed by a tenancy agreement, and although there are many different types of agreement in operation, they do at least set down the expectations that the two parties can have of one another. For example, on the tenant's side, they are usually expected to pay the rent on time and in full, carry out some level of repairs and not cause damage to the property. Landlords are usually expected to keep the building in habitable condition, and allow the tenant to have 'quiet enjoyment' of the property, e.g. not enter the property without giving notice or harass tenants. Tenancy agreements generally set out a relatively neutral set of rules that each are expected to abide by and can be judged against, by a court of law, if necessary.

However, ASB policy has turned this upon its head. Neighbours are now invited to complain about any behaviour they found disruptive or that caused them annoyance. Landlords can now act against tenants based on these complaints – up to and including eviction. In response, Councils and other social landlords built capacity to deal with this, employing an army of anti-social behaviour officers and others to enforce the new rules. Social landlords often issue ASB diaries for tenants to complete recording the behaviour of their neighbours but may also go as far as to issue recording equipment in the case of noise complaints or even install CCTV surveillance.

Contrast this with private tenants and owner-occupiers who are not subject to this same machinery. Private landlords may well be called upon to intervene where their tenants are said to be behaving badly, but they are far from compelled to do so except in some limited circumstance, e.g. drug use on their premises. Likewise, owner occupiers are not subject to virtually any kind of enforcement in such cases. Behaviour may be reported to the police where it is thought to be criminal, but the remedy there lies in criminal charges and the courts, as described above.

This presents problems for proponents of civil liberties and for a system where the rule of law is neutral, power is not arbitrarily exercised. On the first point, the neutrality of the rule of law demands that first, accused persons are held to a neutral standard set down in common law or statute, treated equally by the system and not punished arbitrarily. In the case of the ASB rules, can this be said to be the case?

First, social tenants are held to a standard that other households are not. Their (social) landlords employ staff specifically to monitor and address the behaviour of their tenants and can have highly punitive action taken against them – up to and including eviction (although this must be done via the Courts). Private tenants and owner occupiers are simply not subject to the same rules or surveillance. Second, the definition of ASB is in the minds of those who complain about it. What one person (or people) considers to be ASB may not be so in the eyes of others. This means that what is complained about it one location, may not be in another. The issue of stigma and social housing is relevant here (see CIH, 2020). For example, where social housing estates are stigmatised, this may mean complaints are made about ASB where they may not be in other places, e.g. a loud party on a private estate may be accepted in a way it may not be elsewhere. The effect of this is to hold social housing tenants to an arbitrary and higher standard than others – and this is not something that we should find acceptable. Social landlords also have powers to address ASB that other landlords do not possess, and further, the use of these powers can often be at the discretion of individuals within organisations who are subject to little oversight. This seems in contravention with the principles we have discussed above.

Concluding comments

In this chapter, we have examined the role of the state in modern countries through both a theoretical and a practical lens. The assumptions we make about what theoretically preceded the state may significantly affect our view of its role and the powers that it wields. Most theories of the 'state of nature' assume that as humans we hold several rights purely because we are made in God's image. For Aquinas, these rights were protected by the requirements to act in line with God's reason. The Ten Commandments being the most obvious statement of rights and responsibilities. Even the male, family heads and the monarch were limited by having to act within God's reason.

For Hobbes, individuals must give up their right to do as they please and harm others, this must be reciprocated by all in exchange for the right to live in society and to have their rights protected by an all-powerful sovereign. For Locke, we must all have respect for one another right to live as we wish to, provided we do not infringe upon the rights of others. The role of government, therefore, is to punish those who do infringe upon these rights; but the powers of government must also be limited to prevent exactly the kind of tyranny that Hobbes recommends. Similarly, Nozick recommends a minimal or 'nightwatchman' state that exists only to protect individual rights, and whose powers are limited.

In a housing context, we have looked at several areas of housing policy and how the limitations placed on the state (for good reason) limit its ability to pursue certain policies. We have also seen how some policies pursued by various governments infringe upon the rights enumerated by social contract theories.

This brings us the questions at the beginning of the chapter. What are your thoughts on these in the light of the issues discussed above?

Notes

1 First published in 1651.
2 First published in 1689.
3 First published in 1485.
4 Originally published in 1680.
5 Originally published in 1762.
6 Original publication date is unknown.

Suggested further reading

Irving-Clarke, Y (2019) *Supported Housing: Past, Present, and Future*, Routledge, Abingdon – particularly chapter 3.

Locke, J (2016) *Two Treatise of Government*, Createspace, Scott's Valley CA.

Nozick, R (2001) *Anarchy, State and Utopia*, Wiley-Blackwell, Hoboken NJ.

References

Aquinas, T (2014) *Summa Theologica*, Catholic Way Publishing, Online.

Aristotle (2009) *Politics*, Oxford World Classics, Oxford.

Berlin, I (1969) *Four Essays on Liberty*, Oxford University Press, Oxford.

Carpenter, D (2015) *Magna Carta*, Penguin Classics, London.

CIH (2020) *It's Not Okay: A Guide to Tackling Stigma in Social Housing*, CIH, Coventry.

Cooper, C (2005) Places, folk devils and social policy in P Somerville & N Springings (eds), *Housing and Social Policy: Contemporary Themes and Critical Perspectives*, Routledge, Abingdon, pp.69–102.

Cowan, D & Marsh, A (2018) A perennial problem? On underoccupation in English council housing, *Housing Studies* 34(2) pp.319–337.

Filmer, R (2016) *Patriarcha*, Createspace, Scott's Valley CA.

Fink, H (1981) *Social Philosophy*, Methuen & Co, London.

Garrett, P.M (2007) Sinbin solutions: The pioneer projects for 'problem families' and the forgetfulness of social policy research, *Critical Social Policy* 27(2) pp.203–30.

Habinteg (2011) *Lifetime Homes Design Guide*, Habinteg Housing Association, London.

Heap, V (2016) Putting Victims First? A critique of coalition anti-social behaviour policy, *Critical Social Policy* 36(2) pp.246–264.

Hills, J (2007) *Ends and Means: The Future Roles of Social Housing in England,* CLG, London.

Hobbes, T (2017) *Leviathan*, Penguin Classics, London.

Irving-Clarke, Y (2016) *Supporting People: How Did We Get Here and What Does it Mean for the Future?* Unpublished PhD Thesis, De Montfort University, Leicester.

Irving-Clarke, Y (2019) *Supported Housing: Past, Present, and Future*, Routledge, Abingdon.

King, P (2006) *Choice and the End of Social Housing*, IEA, London.

King, P (2016) *The Principles of Housing*, Routledge, Abingdon.

Locke, J (2016) *Two Treatise of Government*, Createspace, Scott's Valley CA.

Malin, N et al. (1980) *Services for the Mentally Handicapped in Britain*, Croom Helm, London.

Montesquieu, C.L. (2015) *The Spirit of Laws*, Createspace, Scott's Valley, CA.

Murphy, E (1991) *After the Asylums: Community Care for People with Mental Illness*, Faber & Faber, London.

Nozick, R (1984) *Anarchy, State and Utopia*, Wiley-Blackwell, Hoboken NJ.

Rousseau, J.J. (1998) *The Social Contract*, Wordsworth New Editions, Ware.

Scull, A (1993) *The Most Solitary of Afflictions: Madness & Society in Britain 1700–1900*, Yale University Press, New Haven CT.

Squires, P (2008) The Politics of Anti-Social Behaviour, *British Politics* 3(3) pp. 300–323.

Tunstall, R (2015) *The Coalition's Record on Housing: Policy, Spending and Outcomes 2010–2015*, University of York, York.

US (1776) *Declaration of Independence* (online), https://www.archives.gov/founding-docs/declaration-transcript. (Accessed 18 March 2021).

Warburton, N (2010) *Philosophy: The Classics*, Publisher, Place.

3

WHY DOES HOUSING COST MORE, JUST BECAUSE IT'S IN LONDON?

Case study

Conservatives have been in the government in the UK since 2010 (in coalition with the Liberal Democrats between 2010 and 2015). The Conservatives in the UK have a well-known preference for home ownership as the tenure of choice. This is for the reasons I have gone into in some detail in previous chapters.

However, during this time, home ownership rates have been falling whilst the private rented sector has been growing at some pace. The private rented sector has also diversified. It used to be the preserve of young, single adults, couples and students; there are now increasing numbers of families with children and older people living in private rented accommodation.

At the same time, and well before, house prices have been increasing at an exponential rate. According to the Office for National Statistics (ONS), in April 2005, the average UK house price was £155,633; by 2020, this had risen to £278,436, considerably outpacing general inflation.

In London, the increase is even greater. ONS data states that in April 2005, the average house price in London was £235,367; the latest figures from Rightmove[1] show that the most recent assessment of average London house price showed it was £712,707 – nearly a trebling. There are, of course, areas of serious housing need and deprivation in London, and I do not wish to suggest otherwise.

Despite their preference for ownership, the Conservatives have done little to slow the increases, if anything, they have stoked the rise by preferring policies that increase demand; for example, the 'Help to buy' policy

DOI: 10.1201/9781003298212-4

that provides equity loans or guarantees mortgages for first-time buyers. There has been no corresponding attempt to increase supply. The current Conservative government has made promises to 'level up' the country and build 300,000 homes per year.

But, why is this the case?

- Why have house prices increased so much over recent years?
- Why have they increased so much more in London?

Markets

Well, the simple answer is markets. The marketplace exists first, in and of itself, as a place of exchange, but it also serves to indicate the 'price' of goods, and by 'price' I mean the value of a good relative to other goods in the marketplace. Markets can therefore be used as a method of allocating scarce goods or resources at the lowest price possible. There are, however, some prerequisites for it to work effectively. First, everyone participating in the market should have perfect information about what is available and how much it costs. Second, there should be no monopolists in the market, who can control either supply or prices. Third prices must change according to market conditions. Fourthly, players in the market must act rationally; suppliers must act to maximise profits whilst customers act to maximise utility at the lowest cost – a cost/benefit analysis. Lastly, there should be no barriers to entry, anyone should be able to set up as a supplier or participate as a customer. In essence, markets are a method of distributing scarce resources effectively, they encourage competition to drive efficiency (King, 2016).

Although it is hard to think so now, money is not a prerequisite to have a market. The relative value of goods can be represented in other exchanges, for example, historically, barter was the most popular form of transaction. This simply means exchanging one good for another. It might be that I need some of your beef from a cow you have recently slaughtered, and I have recently slaughtered one of my pigs, I might then swap some of my pork for some of your beef. This arrangement does not preclude the relative value being significant either. In a marketplace drowning in beef, I have the power, as a buyer, to negotiate for the best value swap that I can with the multiple beef sellers. When beef is rare, and there is more demand than supply, the seller can dictate a higher value. This type of exchange may also not be so time bound or immediate as in modern trade. It was not unusual for people to agree to share their goods at one point in time and receive goods from others at a later point as a way of spreading the load in the community.

The point I am trying to get to here is that the 'market' is simply a place (not necessarily a physical one) where goods are exchanged for common benefit – only if the trade is advantageous to both parties will it happen. 'Price' then, of a value of one good relative to another, depends on both supply and demand. Greater supply usually results in lower prices, and vice versa. However, supply and demand are not the only variables that matter. Quality is also very important to price. Higher-quality goods are invariably more expensive as they can take longer, and more skill to make; meaning fewer of them on the market and the goods are more durable – the skilled craftsman can therefore demand a higher price. A fourth factor that can make a difference to price is brand. Some brands carry a premium cost because of the prestige that goes with owning their goods, this is to some degree culturally or socially determined. Two items that are very similar may have very different prices, depending on the brand. For example, in the US, both Champagne and Scotch Whisky are relatively expensive, this is because of import costs, but they are also seen as high-status commodities. Quality and brand are two variables that can be applied differently in the housing market – which we will get to.

Some, particularly on the left of politics, often characterise the market as an invention of capitalism to drive profits for owners of capital (as they see it), but as I have explained above, any time there is a group of people making trades for mutual benefit you have a marketplace. Looking at it this way, markets are inevitable. Market arrangements pre-date the institution of capitalism by some several thousand years. There are records and evidence of trade from the times both the Vikings and Romans, Roman soldiers were paid in salt (hence the word salary) instead of money, indeed recent evidence suggest the Vikings were traders par excellence. Great cities have risen and fallen based on trade routes, Istanbul was once one of the world's great trade centres, located at the gateway between the east and west. It's no coincidence that many, if not most, of the world's great cities are located either coastally or on significant water ways. London sits astride the River Thames, Liverpool and Glasgow on the coast and on the Mersey and Clyde, respectively. Istanbul sits on the coast but also on the Bosphorus Strait, Rome on the Tiber, Cairo on the Nile – I could go on. The point being that trade and markets existed in antiquity and long before the existence of western capitalism.

Alternatives to markets

Markets are one method of distributing scarce resources, with price being the signal of relative value, accounting for scarcity (in reality, nothing is unlimited, everything is scarce to some degree), quality and prestige. The only alternative to this is that some authority allocate good and resources, perhaps on their whim or perhaps on some pre-determined basis. If we think back to Chapter 2, before the inception of democratic government, the monarch or his representatives would have had the authority to distribute resources according to their preferences, indeed 'patronage' is still an issue in the present with Governments often seen to grant resources to those it wishes to curry favour with. It's also an issue in the private sector, of course, and there are laws to limit or prohibit such actions for good reason. The best example of this allocation of resources would be rationing of food and clothing etc. during World War II. People were issued with ration books and had to register with shops to collect their ration. Given the circumstances, this was an entirely rational and appropriate response, and indeed lasted for some years after the end of the war for some goods.

Now that we have democratic government, there is an acceptance that some resources (housing included) that should be still allocated or rationed by the government to ensure fairness or justice. Governments, of course, are thought to have the right and duty to do so by virtue of their being elected. This is most obvious in the UK in the National Health Service. Following World War I and particularly as a result of World War II, it was thought unfair that access to medical treatment was available only to those who could afford to pay for it. The NHS (and the wider welfare state) was therefore set up to provide healthcare to all on the basis that those with the means could contribute more to pay for the treatment of those who could not afford to. The state, via the NHS, therefore, must be able to make decisions about who receives what. Just because healthcare is free at the point of need, we must not fall into the trap of thinking that resources are unlimited or that everyone receives the treatment the need/want. There are a number of panels in the NHS to make decisions about the treatment people receive. To take an extreme example, where there is one donated liver, and two potential recipients, a decision must be made about which receives it. How do you decide who gets their live saved? The corollary of this, of course, would be to suggest that it would be the person who could afford to pay more – but that doesn't seem an ethical position either. The point of this is merely to demonstrate that in allocating scarce resources, there is a blunt choice between the market and rationing and sometimes neither can offer a palatable answer.

The other question that this line of thinking raises is who gets to make decisions? In the case above, it seems right that an expert medical panel makes the decision, because they have the knowledge and expertise to assess both patients' health and their long-term prognosis. But, what about other scarce resources? How should they be allocated? How much should they cost?

In former Soviet communist countries, the state did set prices of all goods centrally; it failed because the government could not know all the information it needed to set prices. Each transaction that takes place in the market sends a signal, the thousands or millions of transactions that take place combine to send signals to both suppliers and customers about scarcity, quality, price etc. This is often said to be one of the key strengths of markets over other systems of allocation (rationing). The market does not rely upon any one person or organisation to set prices; knowledge about the state of the market and prices etc. is diffused across the marketplace with each transaction contributing to that knowledge.

Money

As I have said above, money, in the form of coins and notes etc. is not a prerequisite for a market to exist. Indeed, a market based on exchange and barter can quite adequately exist and did so for several millennia. In fact, I suspect such arrangements are still commonplace across the world – even in the West. Money then exists as a medium of exchange and represents the relative value of a good in the marketplace. Using money, it is possible to purchase a good from someone and confer upon them the ability to delay the necessity for them to use whatever you have exchanged. For example, where the exchange involves perishable goods, they must be used fairly immediately, but the use of money enables the sellers to delay their purchase AND decide what they want/need to spend the money on. There are then, some significant advantages to the use of money as a medium of exchange. As I have already suggested, money may not be in the form of notes or coins. Roman soldiers were paid in salt, not only useful in and of itself but also it does not perish so can be used at a much later date. It also has the advantage of being a good that was (and is) universally wanted and used; thus making it an ideal currency.

Money, as we have today, is guaranteed by the central bank of the country that issued it. For example, notes issued by the Bank of England have a promissory note guaranteeing the owner that the bank will redeem the note for the face amount. Notes (obviously) and coins no longer bear any intrinsic value, they are purely representations. However, this was not always the case. Coinage used to be made of precious metals with intrinsic value in the market. The gold coins we are all familiar with from films were actually made of, or contained gold, and had a real value, this value was of course relative to the price of gold or silver etc. But, nevertheless had an actual value. Coinage today has no such intrinsic value. So, in the past, money had a market value attached to the value of the precious metal it was made of. Banks and governments were unable to print money or engage in 'quantitative easing' in the same way as they can now. They could, of course, release more gold into the market thus devaluing it somewhat, but as the supply was limited, could not engage in the printing of billions of pounds in currency that we now see. Moreover, if we go back far enough, there was a time when first, there were no central banks and second, the UK (not uniquely) was made up of several separate 'kingdoms', for example Wessex and Sussex who all issued their own currencies, indeed competed for business (Congdon, 2009). Indeed, banks could also issue their own currency, we see the vestiges of this in Scotland where there are still separate bank notes issues by the Bank of Scotland, Clydesdale bank etc. although all are now centrally backed and have the same value.

Banking

The development of banking may seem somewhat tangential, but banks play a key role in the finance of housing, so a brief discussion of the history of banks is worth undertaking here.

In antiquity, wealth was held and stored based on a physical asset. Mainly in the form of precious metals, e.g. gold. This was often held securely by a specialist such as a goldsmith. The amount of wealth owned and in circulation was therefore based 100% on the value of an asset. At some point, proto bankers realised first that they could issue promissory notes in place of the asset (e.g. physical gold), and second that they could issue notes to more than the value of the asset they own. This relies on those people with promissory notes not wanting their share of the held asset at the same time (a 'run' on the bank). The ratio of bullion to liabilities often fell to well below 100% (Congdon, 2009). Nowadays, the 'safe monetary asset' is no longer gold, but the legal tender notes issued by the central banks (which we will come to shortly).

One of the advantages of banking is that it removes the need for people to trade in physical assets. Initially, as we have discussed, promissory notes could be used to trade in place of a hard asset. Second, the existence of banks enabled the elimination of the need to trade in currency at all. For example, if a person were to trade with another person to the value of £100, instead of taking the notes from bank A, giving them to the other person, who pays them into bank B, this can now be done purely by adjusting the balance sheets of person A and person B with no cash changing hands. The multiplication of this by thousands or millions does not change the principle involved (Congdon, 2009). Likewise, where person A transfers £120 to person B, and person B only £100 to person A, the movement of only £20 on balance sheets is less costly than the movement of £220 in notes. This development was extended to 'clearing banks', where banks could deposit funds and these could be exchanged via balance sheets in a similar way to individual transactions, thus enabling trades of millions of pounds to be conducted in relative simplicity.

Central banking

Despite the long history of banking, central banks are a relatively new feature of the banking system, central banking was unheard of before the 19th century (Congdon, 2009). Indeed, according to Congdon, the history of central banks is somewhat opaque but there is a lively debate about whether central banks appeared spontaneously or were imposed from above by government (see Hayek, 1973). Whatever the origin of central banks, they perform an essential role in the modern world and in housing finance. Central banks are often said to be the ultimate bankers' bank – the safest bank of all. According to Congdon (2009), in the development of banking, those banks that maintained a high asset to notes ratio often became banks for the higher risk banks, thus providing some degree of security. It follows that the bank to the government *must* be the safest bank of all, thus the government's bank and the central bank became one and the same thing. The key issues to the current debate are that first, central banks hold a monopoly on issuing currency, backed by legal restrictions on others; although, we should note that this does not proscribe other forms of trade, e.g. barter. Second, central banks are 'lenders of last resort' in modern finance systems, with responsibility for stabilising these systems.

The individual and the market

As we have established, the market is made of individuals and organisations acting rationally to maximise profit or utility (for the lowest price) in the marketplace. Here, it is worth briefly considering the individual's relationship to the market. If we think back to Chapter 1 (on ideology), I set forward two competing ideas at the base of society, liberalism and socialism/communism (for the moment we'll accept that Conservatism accepts liberal values in this regard).

For liberals, the individual owns his own labour. This is the basis of the right to own property, indeed the ownership of one's own body and labour is the most fundamental form of property rights. There is a whole chapter on property rights, so suffice it to say here, that we own our own labour. In the market then, we also have the right to trade our labour with others for some form of pay. In the modern world, this is usually in the form of cash payment, but in the past other forms of payment have been used, e.g. salt by the Roman army. And, as I have set out above, you can then use this money as a representation of the value of your labour, to exchange for the good you need or want. Generally speaking, the amount you can trade your labour for is commensurate with the level of skill required to carry out the task, the relative abundance of people qualified to do the task at hand, and the value it creates more broadly. Thus, premiership footballers command a very high level of pay because they have a level of skill above many others, indeed the level of skill required to play for a top football side is very rare. Alongside this, it is also true that broadcasters (e.g. Sky) are prepared to pay exorbitant amounts of money to screen football matches, and that fans will pay for tickets and replica shirts etc. thus generating large cash incomes for clubs, shirt producers etc. Likewise, building labourers command a relatively low levels of pay as their job is relatively low-skilled and there is a large pool of people with the ability to do this kind of work.

Socialists or communists take a rather different view. They would say that first the requirement to work to purchase those things essential to life is exploitation of the proletariat by the owners of capital. This exploitation takes the form of extracting the 'surplus value' from the production of the proletariat. To recap, Marx put forward the 'labour theory of value', this states that the value of a good reflects the labour that went into producing it. Thus, where a good is sold and the maker does not receive the full value of the sale price, this is exploitative. The falsity of this is easily demonstrated. How hard a person works is not the sole or defining factor in ascribing value. The value that a particular good can be traded for reflects several factors, and very often not the amount of hard work that went into making them. That is not to say that people who are well rewarded financially do not work hard, of course they do. It merely serves to say that hard work is not the only factor, or indeed a factor at all. If we take the example above, a building labourer carries out backbreaking work, day after day, but because of the low skill level required and abundance of people able to do it, the pay is relatively low. A company chief executive officer (CEO) however, although they work hard, it is not on a level commensurate with physical labour. However, they do perform a highly skilled job that takes many years to qualify for, and there are few people qualified or willing to do what it takes to be qualified – hence they are paid more. Also, a factor in this comparison is the impact of what might happen if the person does not perform their role correctly. For the labourer, the impact will be relatively minimal in most cases; however, the negligence of a CEO may lead to a multi-national company collapsing leading to thousands of job losses, or environmental damage etc.

As well as owning one's own labour and being able to sell it in exchange for money, liberals also accept that people have the right to spend (or exchange) this on whatever they wish in the marketplace. Of course, there are some commodities that are essential to life, e.g. housing, heating, food etc. that we are required to purchase as a prerequisite to living. In the UK (and other countries), there are also some services that we are required to pay for, taxation is levied to pay for all manner of services from the NHS to social housing to running the government and many other things. But we are all entitled to spend our surplus income on whatever we wish, within the law. This means that there is a lively market for goods and services where individuals and organisations compete to sell us these, in order to make a profit. As I have already discussed, in socialist or communist societies, such freedom does not exist. Goods are rationed by the government or those that can be purchased are priced by the government also. Because of this, choice tends to be very limited. One is essentially compelled to buy what the government produces, at the price set by the government.

The state and the market

Thus far, I have been quite negative about the role of the state in the marketplace. Indeed, I have suggested that there is something of a binary situation where the state either leaves well alone and lets the free market operate or sets prices and controls the market almost completely. However, this is assuredly not the case; indeed, the free market cannot exist without the state intervening. If we think back to the chapter on the state and in particular Thomas Hobbes' Leviathan (2014), Hobbes insists on an all-powerful state to prevent people doing whatever they want, including stealing, killing etc. to gain those things that they desire. Although Hobbes' example is an extreme one, it does make the point that some person or organisation is required to prevent crime and punish those that transgress. The same is also true of managing the market.

As discussed above, the market is made up of a multitude of transactions between individuals and organisations for their mutual benefit. Although these transactions are often individual, many transactions rely on longer term agreements or contracts. In housing terms, the most obvious type of contract is a tenancy agreement where landlord and tenant mutually agree to certain duties and responsibilities. For example, as a tenant you agree to pay rent on time and to the full amount, as a landlord you agree to provide a property in good condition, weather tight etc. (although there are problems with this, which we will come to). Housing associations and councils also contract with a variety of external organisations for services, e.g. refuse disposal or building maintenance.

Where individuals or organisations make agreements or contracts with one another, some person or organisation must exist with the power to enforce those agreements and punish parties who do not live up to their duties and responsibilities under the agreements they have made. This is very often the role of the state. In the UK, this role falls to the courts, specifically the civil courts (see Chapter 2). Parties who wish to enforce a contract can take action in the courts to force another party to the contract to carry out a given action they have agreed to. For example, if I take out a loan from a bank and agree to pay it back at a given amount each month, and pay interest on the loan, if I do not do this the bank can take me to court to enforce the terms of the loan agreement. Housing providers may also find themselves in court if they breach the terms of their tenancy agreement, likewise tenants often find themselves in court if they have not paid the rent or behaved in a way contrary to the tenancy agreement. If there were no way to enforce these agreements, they would effectively be worthless. As Jeremy Bentham might have it, they would be 'nonsense on stilts' (see Waldron, 2014). There are other less formal ways in which contracts may be enforced, however. Some agreements, particularly between organisations and governments, have an arbitration process where a third-party panel or tribunal arbitrates on disputes and both parties agree to be bound by their decisions. Such arrangements are common in international trade, for example, or there is an arbitration panel for sportspeople for example. Those taking a more libertarian position would argue that the negative effects of not fulfilling contractual arrangements should be enough to mean that this does not happen. Who would make a contract with someone who has a reputation of not fulfilling what they have agreed to?

The housing market

In the UK, what we call the 'housing market' is both a whole and a sum of a considerable number of constituent parts (King, 2016). First, there are three main tenures, owner occupation (OO), the private rental sector (PRS) and social housing sector (SHS). Although these are often talked about as if they are separate, there are significant social and financial links between them. Housing also has several characteristics that make it unique amongst markets for other commodities.

First, housing is necessary in a way that many other commodities are not. We all need a place to live, to store our other property, for privacy; housing is fundamental to human flourishing (King, 2003). So, of course, is food and drink; however housing has further characteristics that combine with this to make it unique. Second, housing is expensive. As far as purchasing housing, it's probably the most expensive commodity that most of us will ever buy (King, 2016). This means that most people need to pay for it over an extended period of time – probably 30 years or more. Finally, and relatedly, housing is a long-lived asset. Most of our dwellings have been lived in before we did and will be lived in possibly long after we leave it. Houses exist on a timescale sometimes much longer than most people's lives. This has some wider effects. Because housing is a long-lived commodity, mistakes are also long lived. The housing we build today, has implications for future generations. We see this today in the brutalist tower blocks that were constructed in the 1960s and 1970s. These are now highly unpopular and the cause of social problems (Build Better, Build Beautiful Commission, 2020). The present effects of housing constructed some years ago has been thrown into particularly sharp relief by the COVID-19 pandemic as households were forced to stay in the homes most of the time (see Woodfine et al., 2021). Second, because it is a long-lived asset, it can be paid for over an extended period of time. Third, this longevity can mean it falls into disrepair with age and can be very expensive to maintain and repair. Housing requires ongoing investment and in general, the older a house is, the more it costs to maintain in decent condition.

This brings us onto funding. In the past, social housing was largely paid for via taxation and built and managed by local authorities, and private homes largely built by private developers and then ending up in the private rented or owner-occupied sectors. However, the way housing is funded has become much more complex than this over at least the past 40 years. The government now provides significant levels of funding or subsidy across all three sectors (Stephens et al., 2022). So much so that the Chartered Institute of Housing have called for this funding to be rebalanced in favour of more social homes. This means that, in effect, more social homes could be built at zero cost to the government above what they spend now, indeed if these were council homes, they would generate rental income meaning that they would pay for themselves (CIH, 2017). The government then provides funding and subsidies across the sector.

That said, both private and social landlords increasingly borrow money on the private markets, indeed it has been government policy for some years to encourage these landlords to 'sweat their assets' to access private funding sources. Because landlords borrow significant sums in this way, they can be particularly affected by changes in the global money markets and financial situation. Housing is very capital intensive and so often bears the brunt of reductions in spending caused by economic problems. This means that subsidy for housing may end up being reduced at a time when housing affordability is more of a problem for households in all three tenures. Of course, most of the funding for housing comes from private citizens using their earned incomes to borrow money from banks in the form of loans and mortgages to either purchase or improve their homes (King, 2016).

Government intervention in the housing market is generally designed to make housing more affordable – this is the purpose of subsidies. Oxley and Smith (1996, pp.40–41) give a good definition of 'subsidy'.

> *An explicit or implicit flow of funds initiated by government activity which reduces the relative cost of housing production or consumption below what it otherwise might have been.*

This definition raises some germane issues. First, it is neutral, making clear that subsidy can be used for any tenure. Whilst we usually focus on social housing, subsidy can also be given to private housing via housing allowances or improvement grants (King, 2016). Second, this definition does not only refer to the use of public funds. The use of the term 'implicit' funds refers to measures such as rent controls, which effectively provides a subsidy for renters as the landlord cannot increase the rent above a level set by the government. Subsidies can be directed at one group, at the expense of another, or used to build dwellings (production) or to increase affordability for households (consumption).

We should also note here that government intervention in the housing market is often driven by some notion of need or social justice (see Chapter 4). The government usually intervenes to help those in poverty, who are homeless or cannot afford to access housing on the open market – either renting or buying. Since at least 1979, but I would argue earlier, governments in the UK have viewed ownership as an 'aspirational tenure' at the top of what has been described as the 'housing ladder'. This has meant that government has sometimes subsidised households to buy housing when they could have afforded it anyway. This is certainly true of the flagship 'rent to buy' scheme, which certainly subsidised households who could afford to buy anyway, indeed it might even have subsidised people to buy a house that they could not have otherwise afforded without using the scheme. However, this form of subsidy is not necessarily aimed at helping people who could not afford to buy, rather it is designed to promote and encourage a societal value that the government finds congenial to its aims or ideology. Elsewhere in this book, I explain at length why conservatives so value home ownership. Another example of this might be the small tax incentive the coalition gave to married couples. The small amount of the subsidy isn't the point, it would do virtually nothing to help with the overall cost of getting married and raising children. What it does do, is send the message that the government values marriage as the best arena for having and raising children and giving societal stability. Finally, of course, political parties cannot do anything until they are elected to the government, it's not unreasonable therefore, that they use financial mechanisms available to them, to achieve this objective (King, 2016).

Boom and bust

Markets are fundamental to housing (King, 2016). And, markets are fundamentally a method of allocating scarce resources. There are only two ways of doing this, first via the market and second by allocation. This does not mean that the market always works well, or that we don't wish there wasn't a better alternative. The existence of an alternative could be said to be one of the key differences between the political left and right. Conservatives, generally speaking, favour the market as a solution to this problem, whilst the left, for often laudable reasons, favour a greater role for allocation in the system. Allocation allows a greater regard for social justice to be injected into the equation, there is more chance that people will get what they *need* or *deserve*. However, this brings up the question of who needs or deserves priority, and what it is they need to meet their requirements, and how do we decide this?

In the current system we have in the UK, the *Housing Act 1996*, *Homelessness Act 2002* and *Homelessness Reduction Act 2017* set out the framework for these decisions. Local authorities are required to have an allocations policy or framework to allocate social housing to those people deemed to need it, and to exclude other from it on the basis of the criteria the government and local authority. Further the local authority is required to give certain groups of people 'reasonable preference' in their allocations policy. These groups are often ones that we would instinctively agree with such as people who are sleeping on the streets or suffering domestic abuse or harassment. However, it's also true that these policies could be used to give preference to groups that governments find politically congenial. For example, the coalition government legislated to give reasonable preference to people who had served in the armed forces. Now, it is perfectly possible to believe that because of their service to the country, they should be given such preference, but it's also true that there was an element of politics to this. So, if we choose to have a system that operates via allocation, we also must accept that we must make arbitrary decisions about who is eligible for help, because resources are limited.

So, the alternative then is to allocate via the market using the price mechanism. No one argues that price is a particularly fair mechanism from the point of view of social justice or ensuring that everyone has access to the housing they need. However, proponents of markets argue that the market rewards the kind of behaviours that they want to see generally in society. Also, as housing is a store of wealth, and house prices are currently very high, we can see how housing enables those who own it to store and pass on their wealth to their children, when others cannot do this in the same way. As I discussed above, the price mechanism or value of a house (or anything else for that matter) takes in a range of factors, e.g. scarcity, location, quality, condition, whether it needs renovation etc., as the value of housing increases without much reference to these factors, it draws a distinct barrier between those who own housing and therefore have a very valuable and appreciating asset, and those who do not and seemingly stand little chance of doing so.

There is a further factor at play that is worthy of discussion here. Friedman and Friedman (1990) set out four ways of spending money (allocating resources).

1. Spending your own money on yourself
2. Spending your own money on other people
3. Spending other people's money on yourself
4. Spending other people's money on other people.

Further, Hayek states that the most efficient way of spending money is to spend one's own money, on oneself. This is because you have a direct incentive to maximise value (it's your money) and quality, as you will own whatever it is you are buying. This second factor is slightly reduced in number 2, as you will still want to maximise value but not quality in the same way. In number 3, you are likely to want to maximise quality as you will own the good, but as you are not spending your own money, value will not be such a consideration. Finally, spending other people's money on other people is the least efficient way of spending money. This is because there is no incentive to maximise either value or quality. Therefore, the preference should be for people to be able to spend their own money on goods and services for their own use. This is perhaps why those on the political right tend to favour housing allowances that enable households to meet their own needs.

Markets then, often do not always operate in the way that we wish they might. That said, we often project our own priorities onto the market as if it is there to meet them, rather than the reality of market-based system. For example, it is common to hear complaints that the market has failed because everyone, or most people cannot afford to buy a home. But that is not the purpose of the market. When housing is scarce, it will get more expensive, as it does in London. London represents an area that is highly desirable, more people want to live there than there are houses to accommodate them; therefore, house prices are very high. We can also argue of course that house prices in London are too high and being driven by the wrong kinds of purchasers, but the market is doing its job by highly valuing a scarce resource.

Markets can fail though, indeed 'market failure' is often used in relation to housing (King, 2016). The perceived failures of markets often form the main basis for arguments that the government should intervene. Market failure can be characterised as the situation where the conditions do not exist for a market-efficient allocation to take place (King, 2016). This does not mean that they do not work at all, but rather that they do not operate as perfectly as economic theories state that they can. Further, because housing is so valuable and especially vulnerable to changes in government policy, failures can be highly impactful. The 2008 recession provides us with an instructive example.

Although it is easy to do so (particularly so for politicians), the cause of the 2008 financial crash and recession should not be reduced to one causal factor. Instead, there was a multitude of factors that came together in a particular way to cause the crash (Sowell, 2009). The 2008 economic crisis had been building for decades (Blond, 2010).

Blond argues that the commodification of everything coupled with the creation of a global marketplace for capital and overleveraging of governments, banks, families, and individuals led to a situation where debt could not be sustained in the economic downturn. Governments across the western world were forced to 'bail out' the banks by recapitalising them. Many of the banks, which were deemed 'too big to fail', had allowed themselves to get into a situation where they did not have enough reserve capital to meet their debts. This recapitalisation of the banks alongside the economic downturn meant that the UK government (and others) had to take on billions of pounds in debt from the international markets. This is on top of the high levels of debt that the UK government had already incurred as a result of high public spending.

In 1997, Labour inherited an economy in good shape, with national debt under 40% of gross domestic product (GDP). Between 1997 and 2003, they were able to reach a budget surplus as they stuck to Conservative spending plans (as they had promised to, to fend off accusations of profligacy). After 2003, Gordon Brown turned on the spending taps and began to run increasing budgetary deficits between 2003 and 2008, during the biggest boom in GDP on record. This is not necessarily a problem in and of itself, but it did mean that the UK government was already carrying a considerable debt when it was forced to borrow billions of pounds to recapitalise the banks and sustain public spending as tax receipts crashed. Gordon Brown had famously said on several occasions that Labour had ended the 'boom and bust' economy, rapidly amended to having ended 'Tory boom and bust' as the bust took hold. Brown and Labour were particularly keen to emphasise their use of 'Keynesian' economic spending (using government spending to shore up the economy through a downturn) as cover for this borrowing, but this conveniently ignored the core of Keynes' general theory (REF) being 'countercyclical spending'. Meaning that the government should spend counter to economic conditions, e.g. run a budget surplus when tax receipts are high and use these funds to support the economy during recession. Fortunately for Blair and Brown, their rhetoric worked and they were able to portray the economic growth between 2000 and 2006 not as a boom, but as a sustainable pattern brought about by their prudent economic policy.

As King (2010) points out, Brown pursued a policy that was the opposite of what his rhetoric proclaimed. Brown argued for a cautious, prudent approach to public spending to avoid the 'boom and bust' cycle that had typified the UK economy since the 1960s. In reality, he undertook an almost unprecedented expansion in public spending. Health and education particularly benefitted from this. Health spending, for example, tripled between 2001 and 2010 (King, 2010). Brown portrayed these increases as being affordable on a long-term basis. However, personal tax rates remained low, and Brown relied upon a heady mix of increased tax revenues from the financial services sector – possible because of Brown's deliberate policy of light-touch regulation of the City, borrowing and stamp duty from the sale of housing. Blond (2010) and King (2010) agree that housing being treated as a taxable asset was key to the boom and the bust in the late 2000s.

However, Brown's economy turned out to be a one-trick pony which relied upon maintaining economic growth. This was thrown in sharp relief as New Labour got drawn into more and more public borrowing to fund their spending plans without having to raise personal or corporate taxes. There was also a significant increase in personal and corporate indebtedness, as households extended themselves by taking out cheap credit card and mortgage debt, and companies relied on borrowing to fund acquisitions. For example, at the height of the housing boom, there were banks willing to lend up to five times a household's income, on a very low interest rate, on the basis of no more than a self-declaration of that income; no proof required!

The UK thus became increasingly dependent upon borrowing on the international money markets, particularly from China and Middle East (Wolf, 2008). The response of the central banks was to ensure that credit remained cheap via loose monetary policy (Booth, 2009; Congdon, 2009) and maintaining high levels of public consumption became *de jure* if not *de facto* government policy. Rather than the Keynesian model of counter cyclical spending, Brown and New Labour embarked upon its opposite, a massive expansion in government spending funded by borrowing alongside a deliberate policy of keeping taxes low and cost of consumer borrowing cheap.

Alongside the UK being highly indebted and therefore vulnerable to incidents that required high levels of additional borrowing, there was also an international angle to the crisis too. In both the UK and the US, governments had supported owner-occupation over several decades. Ferguson (2008) argues that the promotion of owner occupation as the tenure of choice extends as far back as depression era America, when the housing market was characterised by very short-term mortgages, most of the UK's mortgage market remains on this basis with five-years generally being the maximum fixed term period. Having said that, there is a fortunate group of people whose fixed rate ended, and they dropped onto the variable rate of 1% or 2% above the base rate, meaning that they have been paying under 3% interest on their borrowing for over a decade. Although the base rate is rising again at the time of writing. This also led to changes to mortgage terms for new borrowers where after the fixed rate period ended, mortgages no longer went onto a rate fixed to the base rate, as lenders got stung by the historic low in this rate.

In America, for some time, the government had effectively been underwriting the mortgage market via the quasi-governmental lenders Fannie Mae and Freddie Mac. As happened in the UK later, the Bush (senior) administration pressured lenders not to require full documentation for mortgage applications and Fannie Mae and Freddie Mac were encouraged to underwrite sub-prime mortgages for the first time (Ferguson, 2008). In the UK, albeit later, the Blair government put in place a five-year programme entitled 'Homes for All', aimed at assisting households into owner-occupation. This included a first-time buyer initiative allowing low-income households and keyworkers to buy an equity share in their home, and an initiative called 'Homebuy' that enabled social housing tenants to do the same (King, 2006). As an aside, the government's justification for this policy was not 'choice' as it had been under 'right to buy', but 'social justice' (see Chapter 4). The government's argument was that low-income households *deserved* to own their own home just as much as the more affluent. Whilst this might be true, it is not a good reason to let people take out debts they might not be able to repay. Social justice had long been used as justification for social housing (Brown, 1999), but had now been extended out to ownership, and moreover, to divert resources away from social housing and into home ownership.

The expansion of government debt caused by the existing debt level and recapitalisation of the banks led to the incoming Conservative and Liberal Democrat coalition government having to implement considerable cuts to government spending in order to reduce the budget deficit (the annual increase to the national debt), and eventually the national debt itself. Even the former Labour Chancellor of the Exchequer, Alistair Darling, was warning there would have to be cuts to public spending worse than those under Thatcher. Under the rubric of reducing public spending (or austerity) the coalition government embarked upon a programme of 'austerity', this entailed a whole scale remodelling of the welfare system (see the final chapter) and wider spending cuts to local authority grant funding amongst other areas.

One of the other effects of the crash was that the government and banks became much more risk averse. A decline in house prices at the time did make houses more affordable; for example, in March 2009, house prices had fallen a full 17% compared to a year earlier (Goff, 2009). As I alluded to earlier, interest rates fell to 0.5% in the UK in an attempt to buoy the economy. However, many lenders became much less liberal in their lending, some expecting buyers to pay up to a 30% deposit to buy a home. Birch (2009) found that in September 2007, there were 2,583 mortgage products available to first time buyers and 58% of these products were available at 90/95% of value. However, in March 2009 the number of available products had declined to 1,230 with only 5% of these available at 90/95.

The UK has continued in this vein in the years since. First the coalition then majority Conservative governments have continued with the policy of lauding ownership as the pinnacle of the housing system. Tunstall (2015) argues that the coalition failed to govern the housing system, having given up some of the powers to do so via the localisation of the planning system and budget cuts. Further, the coalition continued the trend of shifting expenditure from object (houses) to subject (people) subsidies, this did not deliver anywhere near the levels of affordable (social) housing needed (Tunstall, 2015).

The current Conservative administration has published a plethora of housing legislation, most of which has been in response to the Grenfell Tower fire and its aftermath. The Social Housing White Paper promised action to resolve the safety issues that had been revealed as contributing to the tragedy. Without going into too much detail here, the tower had been covered in highly flammable cladding which allowed the fire to spread up the outside of the building, negating the fire retardation measures that should have been present within. However, many of these measures were not present. Multiple fire doors were missing, tumescent strips in those doors that were present were old and worn out, fire stops that should have been in place in the internal maintenance channels were also missing or had not been replaced properly. In addition, the tower had a 'stay put' policy in place in the event of a fire. This means that households should have been able to stay safely in their flats, protected by the fire safety measures, until the fire service could get them out. Clearly in the event, this did not happen.

In response, the government has completely overhauled building safety regulation (Building Safety Act, 2022), beefed up the Housing Ombudsman Service to ensure that complaints are dealt with more robustly and is in the process of introducing new consumer regulations on top of the existing governance and viability standards. This includes a new set of satisfaction measures, including a set of tenant perception measures, designed to ask what tenants think about their landlord, housing and neighbourhood.

The most recent government announcement outlined the revivification of the 'right to buy' policy for housing associations. This had been tried under a previous Conservative government who had run a pilot scheme to gauge the level of demand. However, the government was unable to find a way of funding the scheme. Requiring housing associations, who are after all, private bodies to sell their assets raises questions about property rights (see Chapter 5), the government therefore must compensate housing associations. The original plan was to give them the difference between the discounted sale price and market value funded by forcing local authorities to sell their high value assets (buildings) and using these funds; the housing associations would then use these finds to build replacements. This proved very difficult to do and so the policy was dropped. The current government, under Boris Johnson, has revived it but at the time of writing, it is still unclear how it will be funded, beyond guarantees from the treasury that it will be; confirmed in a recent DLUHC Select Committee meeting by the Secretary of State, Michael Gove.

What this shows is that economics and funding are central to housing. Where the government wishes to intervene in the market, it must allocate the funds to do so from its resources. This may mean borrowing money on the international money markets or finding some other way of funding it, for example high value asset sales (as above). Looking back at Chapter 2, it also shows that the state has a key role in the provision of all housing, regardless of tenure. In the social housing sphere, government funds the building of this housing or provides the security necessary for a stable lending market so that landlords can acquire the funds to build. The role of government in maintaining stability in the market should not be underestimated, as it often is. Across the other tenures, the state provides regulation in the form of the planning system and building and safety regulations to ensure that housing fits in with local condition and priorities, and is safe to live in.

So, why does a house cost more, just because it is in London?

As you may have surmised from the above discussion, the answer to this question is not straightforward. There are many factors that contribute to the value (price) of anything. Quality, location, scarcity, desirability may all affect the relative value of a good. For example, a high-quality home, furnished with expensive goods in Mayfair, London, will have a very high purchase price. This is because of the quality of the property and contents, and the high desirability of the location coupled with the scarcity of such properties on the market. We might also suggest that Mayfair is an area where crime is relatively low and there is little to no anti-social behaviour or other social issues. It is also in central London and close to all the social and cultural facilities of the city. These might all contribute to the very high house prices. London is, of course, also in the south of England which is a more desirable location than parts of the north.

Conversely, in the north of England (and other regions), there are areas of depression where house prices are relatively very low. This is because there is often a lack of well paid, or even any, jobs in these areas, so demand to live in them is low. There is also a significant history of neglect of housing and communities in such places, so housing may be in poor condition or surrounded by other empty properties that attracts criminality or anti-social behaviour. House prices in these areas will be relatively low because they may be poor quality or in areas of low housing demand.

Both these examples can get into spirals of price inflation or deflation. High value locations are generally attractive places to live, the more people they attract, the higher prices get; demand increases and property availability decreases. Historically, government policy has been to invest more in these areas as demand is high. However, this only serves to further drive demand. In contrast, low demand locations do not attract investment, employers leave these locations, local shops and services suffer as no one is there to use them and house prices are therefore low. The government does not invest in these areas, and so the spiral continues. However, the current government is in the process of 'levelling up' (Levelling Up and Regeneration Bill, 2022) and has committed to investing significantly across the country to drive investment and regeneration in these areas and communities. The policy of investment in areas of high demand has been dropped in favour of broader spending across the country. There are concerns that this will have the effect of 'levelling down' in London and the southeast as investment moves to other parts of the country, this is important as London is an important source of income to the exchequer, but that said, were house prices to fall in London, that would not at all be a bad thing.

There is one final factor that we should discuss here. The experience of the COVID-19 pandemic seems to have caused some fundamental shifts in the housing market and people attitudes to work, and life more broadly. There are two factors here. First, the general experience of the pandemic, mass furloughing of people from their jobs, being forced to remain in our homes 23 h a day, have caused many people to re-evaluate their lives. Many have left the workforce or changed jobs in what has been called 'the great resignation'. This is combined with a second effect of the pandemic, the rapid adoption of remote working. This was a growing phenomenon beforehand, there was a growing use of laptop computers, cloud-based programmes, e.g. Office 365, and home working, but this was supercharged by the pandemic as many employers were forced to rapidly deploy home working for most or all of their staff. In the housing sector, some organisations estimated they underwent up to five years of IT development and implementation in as little two weeks. Staff were provided with laptops and other mobile solutions, cloud-based systems were implemented and people were equipped to work from their homes. This has meant that homes became offices, and people's patterns of work changed overnight. For some, this freed them from the daily commute allowing them to spend more time with their families or as they otherwise wished. Work time became more flexible as people were no longer tied to offices or other workplaces. People were suddenly able to take breaks in the middle of the day to walk dogs or attend to elderly or infirm relatives, life was transformed and many did not want to lose the gains of this period. There is evidence of people leaving cities and moving to more rural areas as they can now work remotely and home working becoming a 'deal breaker' for many people when they are making choice about work. This has implications for house prices and where local authorities plan to build and what they plan to build. It also has hug economic consequences for cities, as people leave their offices and no longer travel to work. There is a huge network of businesses largely dependent on office workers. Transport systems will need to be rethought as far fewer people commute into cities for work, coffee shops, restaurants, and bars face difficulties as the people they depend on for custom at lunchtimes, and after work, are no longer coming into the cities. The effects of the pandemic will be with us for some time and the housing market will be one of the affected areas as the shakedown continues. We may see house prices decrease in the cities and increase (further) in the suburbs and more rural areas as those people that can work remotely, opt to do so.

Note

1 Used as up to date ONS data was not available at the time of writing.

Suggested further reading

Congdon, T (2009) *Central Banking in a Free Society*, IEA, London.
King, P (2006) *Choice and End of Social Housing*, IEA, London.
Sowell, T (2009) *The Housing Boom and Bust*, Basic Books, New York.

References

Birch, J (2009) No access, *ROOF* 34(3) pp.18.21.
Blond, P (2010) *Red Tory*, Faber, London.

Booth, P (2009) *Verdict on the Crash: Causes and Policy Implications*, Institute for Economic Affairs, London.

Build Better, Build Beautiful Commission (2020) *Living with Beauty: Promoting Health, Well-Being and Sustainable Growth*, Build Better, Build Beautiful Commission, London.

Building Safety Act 2022, HMSO, London.

CIH (2017) *Rethinking Social Housing*, CIH, Coventry.

Congdon, T (2009) *Central Banking in a Free Society*, IEA, London.

Ferguson, N (2008) *The Ascent of Money: A Financial History of the World*, Allen Lane, London.

Friedman, M & Friedman, R (1990) *Free to Choose: A Personal Statement – The Classic Enquiry into the Relationship Between Freedom and Economics*, Mariner Books, Boston MA.

Goff, S (2009) House Price Movements Cause Confusion, *Financial Times* 4/5 April 2009 p.3.

Hayek, F.A. (1973) *Law, Legislation and Liberty vol. 1*, Routledge, Kegan & Paul, London.

Hobbes, T (2014) *Leviathan*, Wordsworth Editions, Stansted.

King, P (2003) Housing as a Freedom Right, *Housing Studies* 18(5) pp.661–672.

King, P (2006) *Choice and End of Social Housing*, IEA, London.

King, P (2010) *Housing Policy Transformed: The Right to Buy and the Desire to Own*, Policy Press, Bristol.

King, P (2016) *The Principles of Housing*, Routledge, Abingdon.

Levelling Up and Regeneration Bill (2022), HMSO, London.

Oxley, M & Smith, J (1996) *Housing: Policy and Rented Housing in Europe*, Spon, London.

Sowell, T (2009) *The Housing Boom and Bust*, Basic Books, New York.

Stephens, M et al. (2022) *UK Housing Review 2022*, CIH, Coventry.

Tunstall, R (2015) *The Coalition's Record on Housing: Policy, spending and outcomes 2010-2015*, University of York, York.

Waldron, J (2014) *Nonsense Upon Stilts: Bentham, Burke and Marx on the Rights of Man*, Routledge, Abingdon.

Woodfine L, Green L, Evans L, Parry-Williams L, Heathcote-Elliott C, Grey CNB, Irving-Clarke Y, Kennedy M, May C, Azam S, & Bellis M.A. (2021) *No Place Like Home? Exploring the Health and Well-being Impact of COVID-19 on Housing and Housing Insecurity. Main Report*, Public Health Wales NHS Trust, Cardiff.

Wolf, M (2009) *Fixing Global Capitalism*, Yale University Press, New Haven CT.

4

WHY ARE THERE HOMELESS PEOPLE AND PEOPLE LIVING IN POOR-QUALITY HOUSING IN A MODERN, RICH COUNTRY LIKE ENGLAND IN 2022?

Case study

The UK is one of the richest, most advanced countries in the world. It has one of the largest, most advanced economies, and a relatively large and inclusive social safety net and health provision system. Yet, homelessness is a persistent and damaging problem for many people. Many individuals and families have no place to call home. Some are sleeping on the streets or 'sofa surfing' where their friends and family will allow them to do so. Others are living in temporary accommodation provided by either the state, housing associations or other, often dubious, landlords. Many households with children are living is such accommodation. Even where people have their own housing, this can be in very poor condition. Some homes in the private rented sector are not large enough for the families that live in them, and be in very poor condition with poor repair, and damp and mould perennial problems. Even in social housing, supposedly in better condition, recent stories on the news have highlighted extremely poor living conditions, with vermin, damp and mould blighting lives. Some families are living in a single room, children to not have space or quiet to do their homework, meaning they do not do as well as they might in education. Damp and mould cause respiratory illness, not to mention the impact of living in a cold home.

- Why is this the case in 2022 (and probably beyond)?
- 'James' presents himself as homeless to the local authority. During the interview he says that he was previously living in Spain and had

DOI: 10.1201/9781003298212-5

a home there that he could have continued to occupy. But, as he is British, he wanted to return to Britain. He came back, deserting his home in Spain and knowing full well he would have nowhere to stay in England.

- What should the local authority's response be?
- What do you think about James' situation from the case study? Have you changed your mind after reading the chapter?
- Is it ever just that anyone is forced to sleep on the streets?
- What if they have been evicted for non-payment of rent?
- What if they have been removed from the home for perpetrating domestic abuse against their partner and children?
- What if they have arrived in this country knowing full well, they have nowhere to stay?

Introduction

As this book concerns the link between philosophy and policy, this chapter draws on material from the opening chapter on ideology and the second chapter on the role of the state. The case I made in my previous book *Supported Housing: Past, Present, and Future* (Irving-Clarke, 2019) was that the manner in which housing for homeless and poorer people is provided is a good proxy for the overall conceptualisation of the relationship between individual and state. This has developed from the highly paternalistic (from the divine right of kings and post-war consensus) to more individualistic social arrangements (Neo-liberalism and the Third Way) as has the nature of provision for those considered vulnerable.

In the modern world, (philosophical) theory and practice are often uncomfortable bedfellows in the housing and homelessness sector (King, 2011), and I would suggest, the welfare sector also. Theory is often seen as an "indulgence that should not be undertaken whilst serious issues such as homelessness persist" (King, 2011, p. 109) or homelessness is often ignored by academics as a practical matter not suitable for theorising. However, Fitzpatrick and Watts (2010; 2018), Fitzpatrick et al. (2014) and Taylor (2018) provide discussions of the nature of rights in relation to provision of housing concluding that it is possible to object to the narrative of a "human right" to housing but still be in support of a legal right to it. Foye (2017) has used Sen's "functionings" approach to discuss housing and happiness arguing against a utilitarian approach and using Sen to translate his own empirical findings into policy recommendations. Taylor (2018) has applied a modified version of Rawls' (1971) difference principle to the *Homelessness Reduction Act (Wales) 2017* concluding that the act does not succeed in making poorest as well off as they can be, given the comparative wealth around them thus delegitimising the Act and related policy. Rather than selecting one philosophical approach and using that, Watts and Fitzpatrick (2018) outline various philosophical approaches to assessing welfare provision and recommend using a "toolkit" approach to analysis – that is the approach I take here. Where theory is applied, it is often consequential in nature; it is applied to a given policy in relation to the outcomes that it produces.

This consequentialism is of course important, and it is common to hear discussions about whether or not a policy delivers *'social justice'*. However, we must also engage with the relationship between the individual and the state, and how the rights of the individual play into policy responses. As we saw in Chapter 1, individuals have certain rights that cannot be infringed by the government, or only after due process. For example, Plato (1971) thought that society should be structured rigidly with all acting according to their talents, children should be removed from their parents at a young age and trained according to their skill-set. This was Plato's attempt to create a genuine meritocracy, where all could prosper by making the most of their talents. Those most suited to philosophical contemplation would become the aforementioned 'philosopher kings', and rule wisely. We would clearly never countenance such a method, parents and children have the right to live together, and family life is seen as sacrosanct. Plato's formulation is intended to promote meritocracy, but clearly makes for tyranny. Such a method has been tried in the authoritarian communist regimes of the 20th century, most notably Mao's China and Pol Pot's Cambodia – hardly models to emulate. The consequence of having a meritocratic system where all make the most of their talents is therefore subordinate to the rights of parents and children to not have the relationship infringed upon by the state.

It is not always easy to divine what 'social justice' or even just plain 'justice' looks like, especially in a societal context where there are multiple competing interests. If we see justice in terms of just one dimension, e.g. meritocracy or happiness, it leads to perverse policy decisions that do not serve the needs of the individual or society. The consequences of a given policy are clearly not enough to justify certain actions. It is therefore important to consider the relationship between the individual and the state when considering what is 'just' and making policies to bring this about.

Housing and social justice

As King (2012) points out, the debate about 'social justice' has tended to focus on the combination of the two words, rather than the idea that the first acts as a qualifier for the second. Critics of the term, such as Hayek (1982) and Nozick (1974), have tried to decouple the two words, arguing that the qualification 'social' seriously disables the notion of justice (King, 2012). In our discussion on making policy about housing and homelessness, we should define what we mean by both 'justice' and 'social justice'; and if outcomes are to be used to assess these, what should those outcomes be?

Miller (1976), in what is perhaps the seminal work on the subject, states that a general definition of justice may be "to each his due" (p. 20). Justice is where we get what is appropriate in the circumstances. The notion of 'dessert' plays a part here, we get what we deserve because of what we have done or the skill and talents that we have (King, 2012). This is different from 'equality', which should be thought of as how our circumstances compare to that of others, rather than how we are judged on our own actions or merits. We can have 'justice' without 'equality', and 'equality' without 'justice', we might even have 'justice' and 'equality' at the same time. Here we should also touch on what is meant by 'equality'. Generally speaking, when we talk about equality, we usually refer to equality of opportunity. A situation where every person can make the most of, and maximise the rewards for, their skills and abilities (as per Miller's definition above). Equality of opportunity can be a tricky concept, as it does not mean that people of equal skill and ability will receive the same reward. As I covered in the chapter on the economy, there are many factors that impact on the rewards that people receive. We might agree that two sportsman who play different games are equally skilled; however, if one is playing football and one netball, the footballer would almost certainly be paid more, not due to a greater skill level, but because many more people are prepared to pay to watch them play, and the football league generates many times more income that that of netball. One could very well make the argument that this is not necessarily fair, but that it is just, under Miller's definition above. The other meaning, equality of outcome, means just what it says, everyone ends up in the same place. This might sound attractive if it means that everyone has the resources and goods they need to meet some social goal; however, achieving it requires such a level of control, on a societal level it is rarely practical.

Defining social justice

Miller (1976) distinguishes between 'justice' and 'social justice'. He defines social justice as

> the distribution of benefits and burdens throughout a society, as it results from the major social institutions – property systems, public organisations, etc.
>
> *(p. 22)*

Rather than a concern with reward and punishment, social justice is concerned with how resources are distributed across society and how they are best allocated. It is concerned with matters such as wage levels, and the allocation of important goods and services, e.g. property, rights, education and healthcare. In terms of housing policy, 'justice' might lead us to a position where we defend property rights at all costs, where a concern for 'social justice' would lead us to a discussion about the distribution of housing and whether all have access to good quality housing at a price they can reasonably afford. This is especially relevant in housing markets like the UK where home ownership is often prioritised above social provision. If social justice is to be determined by the consequences of a given set of policies, then we need to decide what outcome(s) we can use to make this determination. Several social theorists have attempted to offer a definition of social justice and the outcomes that might be used to assess whether, or not, a given policy delivers it.

Jeremy Bentham (REF) made such an attempt with 'utilitarianism'. His theory states that the most important outcome in making a policy decision is that it maximises aggregate happiness. Not the happiness of any individual or any one group, but the aggregate sum of happiness in society. Making everyone happier sounds like a reasonable goal, but it also may lead us to make some questionable decisions. In the famous example, if a train is running out of control, will hit another stationary train, and the only way to prevent this is for a person to lie on the track, diverting the train over the points onto a parallel track but being killed in the process, the happiness of the many people whose lives are saved on the stationary train outweigh the grief caused by a single death. Bentham would therefore find it justifiable for state policy to coerce a person to lie on the tracks – increasing aggregate happiness being the only acceptable policy outcome. Clearly, this is a suggestion we would recoil from. The state (rightly) does not have the power to compel the death of a citizen in these circumstances, regardless of how happy it might make others. More relevant here, we might ask what this person has done to deserve being coerced into death, how does this action serve justice for them? This leads to the conclusion that the increasing of aggregate, or even individual, happiness cannot therefore be a good outcome on which to base notions of justice.

A theory of justice

In *A Theory of Justice*, John Rawls (1971) invites us to consider what a just society might look like. He does this via a thought experiment where we are asked to stand behind a 'veil of ignorance' that removes our knowledge of our own gender, employment status, intelligence, aptitudes etc.; whilst leaving us with knowledge of politics, economics, the basis of human organisation and human psychology. From, what he terms, this 'original position', we should set out the general principles upon which we would want society to be organised to deliver social justice. His theory being that as we do not know our position in society from this 'original position', we will choose to structure society in such a way as to maximise the benefits for all, whilst least compromising the generation of the wealth that provides those benefits. In making this judgement, Rawls sets out two principles, one based on liberty, the other on distribution of goods/wealth.

The liberty principle

Here Rawls states that from behind the veil of ignorance, we would want everyone to have the same basic liberties compatible with ensuring liberty for all. This means that you would have the right to speak freely, freedom of conscience, religion and association – the rights we are generally familiar with. However, these rights would end where they curtail the rights of others. The intervention of the state via the rule of law is justified in these circumstances to prevent the rights of others being violated. As Rawls orders these rights lexically, this must first be satisfied before moving to the second of Rawls' principles.

Equality of opportunity and the difference principle

Second, Rawls states that inequality related to certain offices or jobs can only exist where these jobs are open to all, that is to say that none are excluded on spurious grounds, e.g. age, sexual preferences etc. Rawls also goes further than this including unfettered access to the provision of education and training to allow people to meet their potential. The second part of Rawls conception is the 'difference principle'. This states that inequality is only acceptable in circumstances where the worst off are in as good a position as they can be. Warburton (2014) uses the example of income to illustrate this. If we must choose between a position where a few earn high wages but the lowest 10% of earners barely earn enough to survive; or where the standard of living is generally lower, high earners earn less but the poorest have a good standard of living; from the original position we would choose the latter. From Rawls' point of view, it is very much outcomes that matter, and state intervention is not only required, but essential to ensuring outcomes meet the objectives he sets out in his *liberty* and *difference* principles.

Rawls differs from Bentham in that his metric for 'social justice' is not increasing the aggregate amount of happiness but maximising the well-being of the poorest. Rawls' ideas too, sound reasonable. Who would not want the poor to be as well off as they can be? However, Rawls' ideas have been criticised, not least by his colleague Robert Nozick (1974) who presents a very different, yet no less compelling, version of justice.

Nozick, Anarchy, State and Utopia

In *Anarchy, State and Utopia* Nozick (1974) presents his view that *a priori*, the rights of the individual are paramount. The individual has rights that are inviolable by either other individuals or the state, they have ownership (property rights) (see Chapter 3) over themselves in the same way that they have ownership over inanimate objects; therefore, inviolable right of control or transfer over themselves and their assets. Nozick tests his ideas to destruction. Whereas Plato tests the idea of meritocracy, showing us what abuses are necessary to genuinely have this; Nozick shows what necessarily happens if individual liberty is prized above all else.

This position leads Nozick to state that there is no such thing as the public good (in contrast to Mill (Mill & Bentham, 1987) and Rawls [1971]), only the benefits or disbenefits accrued to the individual (King, 2010). Nozick puts it thus,

> there is no **social entity** with a good that undergoes some sacrifice for its own good. There are only individual people, different individual people,

> with their own individual lives. Using one of these people for the benefit of others, uses him and benefits the others. Nothing more. What happens is that something is done to him for the sake of others. Talk of an overall social good covers this up. (Intentionally?) To use a person in this way does not sufficiently respect and take account of the fact that he is a separate person, that his is the only life he has. **He** does not get some over-balancing good from his sacrifice, and no one is entitled to force this upon him – least of all a state or government that claims his allegiance (as other individuals do not) and that therefore scrupulously must be **neutral** between its citizens.
>
> *(Nozick, 1974, pp. 32–33 [author's emphasis])*

Ownership looms large in Nozick's writing (Wolff, 1991). Nozick's view that the individual has inviolable rights (sovereignty) over himself and his property, his belief that government should be neutral leads him to assert that state action to benefit one individual at the cost of another is unacceptable. This is a use of the second formulation of Kant's Categorical Imperative (Kant, 1997) that an individual must only ever be treated as an end in themselves and never a means to the ends of others. What this means is that one cannot formulate some utopian end state (as Marx and to some degree, Bentham and Rawls do) where one person's resources are commandeered for the benefit of others (King, 2011).

Instead, Nozick asserts that it is the legitimacy of the process of exchange that matters; where an individual accrues assets by a fair means of exchange, he has an inviolable right to use these assets as he pleases. Where the accrual of assets has not been via fair means, Nozick finds that it is a legitimate role of the (minimal) state to mediate and order reparations thus in the case of the trans-atlantic slave trade, for example, there are accruals to some individuals that are the result of unfair (to say the least) practices (i.e. to say outside fair contracts or exchange). Slavery of course also first violates the second formulation of the categorical imperative (as above) and another Nozickian concept, 'side-constraints'. This concept states that individual freedom is inviolable only up to the point that it infringes on that of another; best explained by the famous maxim "My right to swing my fist stops where your face begins!". For Nozick, individual rights of property (including mastery over oneself) are inviolable but only to the point that first, one cannot use another as means to any end; and second that the sovereignty of others isn't violated. In these, and only these, circumstances can intervention by the state be justifiable, to right the wrong and provide reparations (compensation). Nozick's concept of 'social justice' is concerned with *process*, whereas Bentham and Rawls' are concerned with *outcome*.

Rawls' accepts that inequality exists but also insists that the poorest must be as well off as they can be for this to be perceived as a just situation. The state's intervention to ensure this is the case is entirely justified for Rawls, for example, the taxation of the wealthy, in order to redistribute wealth to the poor is desirable, even essential for Rawls – even though it entails using some people as a means to the ends of fairness, and the welfare of others. For Nozick, this is completely unacceptable. Taxation and redistribution violate the sovereignty of the self and of private property (see Chapter 4), uses some as a means to the ends of others and violates the concept of side-constraints. All state action is backed by the threat of force, for Nozick, the use of force in this way is unacceptable.

Sen, Nussbaum and 'capabilities'

In response, at least in part, to Bentham (Mill & Bentham, 1987), Rawls (1971) and Nozick (1974), the economist philosopher Amartya Sen (1974; 1974a) criticises these approaches as focusing on the 'happiness' or amount of resources an individual has. Instead, he focuses on the ability of the individual to achieve well-being as a function of what they are able to do and be, and thus the kind of life they are able to lead. Sen focuses on the things people are able to do (doings) and what they are able to be (beings) as a result of the resources available to them. He calls these notions, 'capabilities'. These are the actual freedoms one has to achieve one's potential – not merely the freedom or right to do so, but the substantial opportunity to achieve it. This, says Sen, is actual freedom.

Thus, is his essay, 'Equality of What?' (Sen, 1979a), Sen asks us to consider two persons with the same resources but one of whom has a disability. The person with the disability may not be able to use their resources in the same way or as widely as the able-bodied person. For example, if they are a wheelchair user, they may not be able to access many buildings or services in the same way as the able-bodied person; or may have to expend some of their limited resources on adjustments because of their disability. They may need to purchase a wheelchair for example or require taxis where public transport is inaccessible. It's easy to see in this situation, therefore, that their opportunities are more limited than for the able-bodied person. Sen also criticises utilitarianism as being too subjective, and as potentially, easily leading to a situation where more resources are expended on the able-bodied, where this leads to greater levels of happiness. Despite setting down this framework, Sen stops short of providing a list of actual 'capabilities' as those that are core to well-being. However, Martha Nussbaum (2011) has provided such a list, as follows:

1. *Life* – being able to live to a 'normal human life length'; not dying prematurely
2. *Bodily health* – being able to have good bodily health, including reproductive health, to be well nourished and have adequate shelter
3. *Bodily integrity* – freedom of movement and freedom from assault or domestic abuse; have opportunities for sexual satisfaction and reproduction
4. *Senses* – being able to use the senses, to imagine and think in an informed way and as a result of an adequate education, freedom of expression
5. *Emotions* – being able to have emotional attachments to things and people outside ourselves; freedom to feel, love, grief, longing and justified anger
6. *Practical reason* – being able to form a perception of 'the good' and plan one's life accordingly; freedom of conscience

7. *Affiliation* – the ability to recognise and show concern for others, to engage in social interaction and empathise with others; freedom of assembly and political speech; non-discrimination against you
8. *Other species* – being able to live with and show concerns for other animals, plants and the natural world
9. *Play* – being able to laugh, play and enjoy recreation
10. *Control of one's environment* – being able to participate in political choices, exercise political choice, freedom of speech and association; being able to hold property and having equal property rights as others; freedom from unwarranted search or incarceration.

Although others have taken Sen and Nussbaum's work and developed it considerably (see Crocker, 2008; Alkire, 2002; Wolff & de-Shalit, 2007) for example, for our current purpose, we can confine ourselves to the basic framework set out by the former two thinkers. Much of Nussbaum's (2011) list of 'capabilities' is familiar, the right to free speech, assembly, conscience etc. are all basic rights from Locke (2016) especially. However, the 'capabilities' approach asks us to consider what resources individuals may need to realise these rights as actions, rather than merely conferring the right to not have them infringed by others, particularly the government. In this sense they are positive rights – some action by the state or others is required to support some people to secure their rights. The 'right to housing', if such a right exists, is explicit in Nussbaum's (2011) list not least in number two, which includes the right to 'adequate shelter'. Along a similar vein, King (2003) argues that housing is fundamental to human flourishing and should be seen on a par with the right to property. King argues that housing provides a venue where people can perform certain functions, e.g. toileting that cannot be performed in public. Housing therefore provides the privacy required for a 'good' and dignified life.

Housing policy and social justice

Housing has been a key part of the welfare state, and therefore the social contract, for centuries in the UK (see Irving-Clarke, 2019); more recently it has also been key to delivering 'social justice'. In my previous book, *Supported Housing: Past, Present and Future*, I argued that the provision of housing in the UK has largely been ideologically driven, I continue that analysis here.

As I discussed in Chapter 1, prior to the 'enlightenment' and the Glorious Revolution of 1688, that introduced parliamentary sovereignty, the UK was an absolute monarchy. The thesis of the divine right of kings (see Aquinas, 2014; Filmer, 2016) held that the monarch was placed on the throne by God and therefore held the 'divine right' to rule over his/her subjects as he/she saw fit. Common people were viewed as the 'subjects' of monarchy, often the figurative children of the sovereign, entrusted by God to their care. The wider system of governance was feudal in nature, most of the land owned by the monarch or feudal 'lords of the manor' with the majority of people subject to the rule of these literal 'landlords'. This led to a highly paternalistic system of support with what support was available being in the gift of the monarch or their representative – who could be tyrannical or benign as they wished. This support frequently took the form of measures that were hardly supportive by today's standards, for example, being put in the stocks for three days and removed forcibly from the parish was a standard policy response. These measures of 'hard paternalism' have been reflected more recently in the work of Lawrence Mead (an American conservative). Mead (1992) contends that welfare recipients are not self-regarding, rational actors but require stern action from a tutelary state to move from welfare to work (or at least off welfare) in order meet society's normative expectations.

For much of history, support has been far from socially just, supportive or what we think of as even humane today. The 16th-century 'poor laws' were a collection of local and national measures that attempted to provide a basic welfare safety net. The Act for the Relief of the Poor 1597 (amended in 1601) established the first national framework for relieving poverty by establishing 'Overseers of the Poor' responsible for distributing food, money and other essentials to those in poverty. The 1601 amendment enabled the creation of alms-houses or poor-houses for the 'impotent poor', those who could not help themselves. Such assistance from the state came at a price. The able-bodied poor were set to work in a 'House of Industry' where materials were provided for the poor to work productively, the idle poor (presumably those who refused to work in a House of Industry) were sent to a House of Correction or even prison whilst pauper children would become apprentices. This system continued until the 19th century.

The system for assisting those in poverty was updated by the *Poor Law Amendment Act (PLAA) 1834*, provision was placed under a single system of administration (rather than the existing parish-based system). A Poor Law Commission was established to oversee the operation of the new system at a national level. Parishes were clustered into Poor Law Unions (each run by its own Poor Law Board), and Union Workhouses established to provide relief to the poor in the area covered by each union (Irving-Clarke, 2019). At this time, relief to the poor was provided both as 'outside relief', financial or charitable relief provided direct to the poor, and "inside relief" – the workhouse. This was a place where the poor were provided with unpaid work, which paid for their board and lodgings. During the 1840s, the workhouse became the only avenue of relief to the poor (Irving-Clarke, 2019). Workhouses did not distinguish between the circumstances of paupers and were designed to act as a deterrent to poverty through a reputation of brutality, but as Crowther (1981) notes, the reality was often more one of neglect and boredom. There are clear echoes of 'utilitarianism' is this approach, getting the poor out of sight of society so they don't have to think about them.

The poor law system continued well into the 20th century and homelessness policy failed to keep pace with positive changes to wider housing policy after World War I. For example, the Tudor-Walters report highlighted the shortage of good-quality homes for men returning from the war and recommended a huge housebuilding programme as well as introducing space standards (Park, 2017). This was implemented via the Housing and Town Planning Act 1919 (the Addison Act). Provision for homeless people continued to consist of large, poor-quality converted workhouses and resettlement units, known as 'spikes' with shared dormitories, they were predominantly associated with single, white, middle aged 'men of the road' with alcohol problems (Neale, 1997). Despite the National Assistance Act 1948, hostel provision continued to be large and basic, some having up to 100 beds primarily used for emergencies and for those excluded from accessing other accommodation (Busch-Geertsema & Sahlin, 2007).

Until the early 20th century, welfare provision in England was based in a liberal conception of the role of state, an acceptance that the state should have a minimal role in intervening in social conditions, and that first people of means have the right to use their wealth as they wish to, and a Nozickean belief that wealth fairly acquired was not to be subject to state meddling. This was couple with the view that, as per Aquinas (see Chapter 1), people's 'station' in life was pre-ordained by God, and attempts to change this would be to interfere in a divine plan.

This began to change in the early 20th century with Octavia Hill's efforts around the London slums. Her overall aim was to make 'lives noble, homes happy and family life good'. Although her moralistic and paternalistic approach was that properties should be let to tenants at a high standard and in return, tenants would be expected to 'moderate' (change) their behaviour. This should be seen as a continuation of the 'poor law' approach to poverty. During the 'poor law', the state played a minimal role in alleviating all but extreme poverty, wealthy merchants, the church and landowners often voluntarily took action to do so via the provision of monetary or other donations, or in the case of the church in particular, the establishment of almshouses (see Pannell & Thomas, 2000; Irving-Clarke, 2019). Indeed, the concept of *noblesse oblige* compelled those with means to provide assistance to the poor on a moral level. Around this time, some wealthy factory owners also built housing for their workers to live in, the best known of these is probably Bourneville Village, built by the Cadbury family for workers. This was a tacit recognition that a healthy workforce is a happier and more productive workforce. Provision of better quality housing was well and truly on the policy agenda. The *Housing and Working Classes Act 1890* empowered local authorities in London to build housing for poorer communities and charge reasonable, fair rents. This was the first time the state had played a role in the provision of housing to a noticeable extent. Councils had to re-house a least half of the people displaced by these slum clearances. A decade later the *Housing of the Working Classes Act 1900* extended the 1890 Act to areas outside London and allowed councils to buy land outside their own districts.

At the start of World War I nearly 90% of households lived in the private rented sector, with most of the remainder being owner occupiers (King, 2006). The outbreak of the war led to large-scale demographic change as working aged men were conscripted to fight, leaving significant gaps in the labour force.

The need to produce munitions and other war supplies meant that those left (mainly women) moved *en masse* to cities to work in factories. The progress of the housing the working classes acts was lost as construction of new homes stalled and materials were directed to the war effort – this led to a shortage of urban housing and as a result rents increased dramatically. This led to several 'rent strikes' in areas such as Glasgow (whose shipyards were vital to the war effort), in response the government introduced fixed rents and mortgage interest at 1914 levels via the *Rent and Mortgage Interests (War Restrictions Act) 1915*. The housing shortage was only made worse after the war ended as soldiers returned and started families (King, 2006). The idea that every household should have a decent place to live was therefore predicated on the war effort and the suffering on the front lines, rather than any notion that decent housing should be a right in and of itself. The tail end of World War I also coincided with the Russian Revolution in 1917, the ideas of Karl Marx and Friedrich Engels (see Chapter 1) were being given life by revolutionary movements, and better conditions for the working classes and poorer members of society were seen a way of placating a revolution (Ravetz, 2001). This led to the implementation of '*Homes fit for Heroes*' (see Yorke, 2017) via the *Housing and Town Planning Act 1919* which introduced a national housing programme that highlighted the shortage of good quality homes for men returning from World War I, and the surge in new households.

There was also acceptance, for the first time, that private enterprise would not be able to produce enough quality homes at affordable prices and that the state had a role to play. The idea that the market could not provide homes for all led to a 'safety net' provided by the state to provide for those who were unable to access the market. However, the interwar years also saw an explosion in owner-occupation as wages rose in real terms, house prices remained static and the creation of building societies with high levels of liquidity allowed a stable, secure base for investment. Thus, the sector with the highest growth between the wars was owner-occupiers, and at a time of cuts to local authority grants.

A brief history of social housing

Social housing has a relatively short history in the UK with much of it being constructed as part of the post-war welfare state (c.1945–1979). The country's experience after the first war, invigoration of Marxist ideas about socialism and communism, a genuine determination to improve the wellbeing of the population, and a two-pronged problem of a housing shortage (caused by bomb damage) and returning soldiers and the related 'baby-boom' and growth in household formation (Irving-Clarke, 2016; 2019), drove the introduction of the Beveridgean welfare state. Once again, the role that people had played in winning a war loomed large in housing policy, larger than any notion of decent housing being a right, in and of itself. That said, it can be argued that housing, and the wider welfare state, were a just reward for not one, but two catastrophic wars against tyranny.

Housing became a pillar, albeit a wobbly one (Torgerson, 1987) of post-war rebuilding and delivering social justice. The Beveridge Report (1942) introduced the concept of a comprehensive state safety net to protect its citizens from the five giant evils in society (squalor, ignorance, want, idleness, and disease) from cradle to grave. With the words of Beveridge ringing in its ears the Labour government, led by Clement Atlee, chose to meet the demand for new homes mainly via large scale local authority building programmes. House building numbers remained at over 150,000 units per annum right through the 1950s and were still above 80,000 per annum throughout the 1970s. It was in good financial shape with many older council properties were free of debt enabling reductions in rent or used to subsidise further building (King, 2006).

By 1971, councils housed almost one in three households (and would do so until the early 1980s), owner occupation was also growing and fewer than one in five households were privately renting – this number would continue to fall. In addition, two-thirds of working age social-renters were in full time jobs, the broad tenant base and strong position of social housing could have been a springboard for further expansion (Stephens et al., 2018). However, Tucker (1966) argues that the high-volume building for slum clearance did not have the earlier post-war commitment on building quality homes resulting in both the standard and status of social housing falling. At the same time, the link between housing and its impact on all areas of life was established. Housing was promoted as a crucial political issue; the manifesto on which the Conservatives fought the 1951 election, which stated:

> Housing is the first of the social services. It is also one of the keys to increased productivity. Work, family life, health and education are all undermined by crowded houses. Therefore, a Conservative and Unionist Government will give housing a priority second only to national defence.
>
> *(Conservative Party, 1951)*

Despite seeing housing as second only to national defence at the 1951 General Election, the Conservative government of Harold Macmillan (1957–1963) was the first to challenge the 'post-war consensus' by questioning the mass provision of housing by the state. Housing is often cited as 'the wobbly pillar' under the welfare state (Torgerson, 1987, pp. 116–126), because of the partial extent in which it has been considered as a responsibility of the government and of social policy.

The ideological debate about the state's role in mass housing provision and the encouragement of those who could do so to move out of social housing was twinned with the fact that social housing was becoming less easily available. Local authorities were encouraged to allocate council housing on *need* rather than *desert* following the Cullingworth Report (1969) meaning that the essence of social housing was changing. Mullins and Murie (2006) point out that council housing had always been provided to those who could afford the rent, e.g. the more affluent working class and that those who could not afford to pay were residualised in the sub-standard private rented sector (or worse). In 1979, total 20% of families in the top decile of income distribution lived in social housing, compared to almost zero by 2004/05 (Mullins & Murie, 2006).

Prior to the 1970s council, rents were directly linked to the actual cost of provision (as opposed to being market driven), there were no 'subject subsidies', e.g. compulsory rent rebates or housing benefit to assist with housing costs (*ibid*). These were not introduced until the early 1970s (King, 2006). The introduction of compulsory rent rebates in the *Housing Finance Act 1972* broke the link between cost of provision and rents charged and began the residualisation of the poorest people in council housing by removing the cost of a social rent as a disqualifying factor. Residualisation is defined by Malpass and Murie (1982, p. 174) as,

> The process whereby public housing [and other social housing] moves towards a position in which it provides only a 'safety net' for those who for reasons of poverty, age or infirmity cannot obtain suitable accommodation in the private sector.

Until this point, council housing had been an aspirational tenure for the relatively wealthy working class (CIH, 2018). Prior to the 1972 Act, local authorities did have the power to charge a subsidised rent on a discretionary basis but as Cowan and Marsh (2018) point out, the lack of security of tenure (only introduced in Right to Buy legislation during the 1980s) left households vulnerable to eviction at the whim of local authorities; and withdrawal of subsidised rent could be used as a lever to produce a desired result, e.g. moving to a smaller property in cases of under-occupation (CHAC, 1953, p. 14). Although Cowan and Marsh note that this was not common practice.

The *Housing (Homeless Persons) Act 1977* continued the residualisation of the poorest people in council housing, local authorities now had a duty to house people considered vulnerable under the law and provide financial assistance to help them afford the rents. Council housing was thus used to discharge this duty and the introduction of compulsory rent rebates (subsequently Housing Benefit and now the housing element in Universal Credit) meant that, as stated previously, affordability was no longer a disqualifying factor.

The election of a Conservative government under Prime Minister Margaret Thatcher in 1979 altered the course of housing policy permanently. She is often largely blamed for the current housing situation in the UK, but the residualisation of council housing was already in train from as early as the 1960s, indeed Malpass and Victory (2010) find that this residualisation was implicit in, and a result of, housing policy throughout the 20th century. Thatcher's conservative social views and adherence to the monetarist policies of Milton Friedman which prescribed limited state intervention in a 'self-correcting' free market drove housing policy during the 1980s and well beyond (Irving-Clarke, 2016; 2019). Gough (1980) argues an ideological attack on welfare was firmly cemented by the Thatcher governments; although he suggests this was largely initiated by the Labour governments of Wilson and Callaghan in 1975. Henderson (2019) argues that the qualitative shifts in social policy were aimed at reasserting individualism, promoting self-reliance and family responsibility, and to ultimately dismantle the ideology of the collective social provision of the post-war era. Thatcher's Conservatives focused on setting people free from government constraints on their lives by rolling back the frontiers of the state. There was a re-emphasising of a commitment to market provision of housing, promoting a revival of the private rented sector, and a belief in extending home ownership with social housing being aimed only at those who could not compete in the market. Clarke et al. (2000) and Gough (1980) argue separately that this was a permanent revolution impacting on the scale, forms and social relationships of welfare; the emergence of neo-liberal policy represented a sea change in the ideological approach to welfare and housing.

Driving policy was the government's preference for home ownership over other tenures, this is depicted in the rhetoric of the *Housing Act 1980* (see Gurney, 1996) which introduced the controversial, flagship 'Right to Buy' policy. Local authority tenants could now purchase their property at a discount dependent on how long they had been a tenant. The policy proved extremely popular with voters and has never been fully repealed by successive governments of either party in England; although the Housing Act 2004 introduced stricter regulations including a tighter time frame in which tenants could sell after purchase and with less discount to buy. Indeed, King (2006) argues that 'right to buy' is one of the most successful housing policies ever, in terms of its own policy objectives. The Conservative governments of the 1980s and 1990s continued a doctrinaire repositioning of state-controlled housing out of the government arena via the *Housing and Planning Act 1988* which gave councils the option of transferring all or part of their housing to another landlord, such as a Registered Social Landlord (RSL). The 1988 Act allowed council tenants (as a whole) to choose to transfer their existing homes to another landlord, either a separate housing association (large-scale voluntary transfers, LSVT) or arms-length management organisation (ALMO) operated by the local authority but managed autonomously.

The return of (New) Labour

The New Labour governments (1997–2010) followed a Conservative, neo-liberal agenda in some regards, particularly in relation to housing (see King, 2006a). However, Newman (2001) finds that (New) Labour also "attempted to establish – and sustain – a new set of political alliances. It sought to forge a consensus around an agenda of 'modernising' reforms designed to remedy deep-seated social problems" (p. 1). Newman also found a softening of the ideological commitment to market mechanisms and competition as drivers of public sector reform. There was a reimagining of the cultural and social spheres as New Labour attempted to retain the economic gains of the previous Conservatives governments whilst reshaping civil society via the invocation of moral and civic values and a new emphasis on citizenship and social inclusion. They also sought to appeal to new groups, women, black and minority ethnic communities, disabled people, LGBTQ+ groups and the young; New Labour saw these were groups as excluded from the political system. At the same time, they sought to play down class-based enmity – a corollary of 'Old Labour' (Newman, 2001). For example, women were expected to become "full worker-citizens" (Newman, 2014, p. 442).

The Third Way can be defined against the two prevailing ideologies of the 20th century. The 'first way' of the 'old' left emphasised social justice via a universal welfare state, Keynesian management of the economy to provide full employment and using the state to accomplish social policy goals, whilst also acknowledging the role of the private sector in a mixed economy. The 'second way' of the New Right favoured a laissez-faire approach to the economy, finding free-market solutions to all problems, and placed emphasis on social justice as personal freedom and economic gain, over social issues and values (Driver & Martell, 1999). The 'Third Way' embodied elements of both previous systems, promoting the market alongside social justice, and community, embracing market enterprise, without automatically favouring market solutions, promoting a positive role for the state but did not assume the state should provide services directly; and offered a communitarian view of society rather than an individualistic one.

New Labour wanted to preserve a core concern with social justice whilst accepting that the range of issues that divided left from right was shrinking. Where equality was seen to conflict with individual autonomy, new relationships between the individual and society must be forged (Giddens, 1998). Giddens (the intellectual guru of the Third Way) distils this down into two basic maxims, "no rights without responsibility" and "no authority without democracy" (Giddens, 1998). These can be seen as the Third Way conception of 'social justice'.

No rights without responsibility

Locke (2016) sees that certain rights that are ours as a result of merely being human (life, liberty and the pursuit of happiness, as in the US constitution) and the role of the state is to preserve these rights and to ensure they are not infringed upon. Giddens (1998) proposed that decreasing social solidarity and increasing individualism demanded a new contract, proposing that rights can only be granted where corresponding responsibilities are also adhered to. He is also clear that such conditionality must apply to all in society, not only those in need of state assistance (as the right often propose).

No authority without democracy

In direct contravention of a Conservatism that looks at tradition as a means of conferring authority on institutions, e.g. the church, family or state; and without due deference to these institutions, authority crumbles (Scruton, 2001; 2006), Giddens argues that increasing individualism and decreasing faith in, and attention to, custom and tradition mean that (again) a new social agreement is required. For Giddens, the sole source of authority should be via democracy.

These two concepts underlie Giddens, and therefore New Labour's vision of the "social investment state".

Giddens is critical of the neo-liberal conception of society as a meritocracy, one where the conception of equality is that of equality of opportunity. As far as he is concerned, this concept of equality leads to unacceptable inequalities and some people and groups receiving rewards that are disproportionate to their talents or contributions. He argues that this inequality leads to some groups in society becoming dislocated from the rest, at the top and the bottom. Extreme wealth at the top allows a group of 'untouchables' to pull away from society and choose to withdraw from social institutions. At the bottom, the very poor are rendered unable to participate in civil society because of lack of resources.

The Third Way conception of equality is as "equality as inclusion" and this is the key to the Third Way conception of the relationship between the individual and the state. Giddens (1998) argument is that where state interventions benefit most of the people living in a society, this fosters a "common morality of citizenship" (p. 108); whereas when there is a purely negative connotation of welfare, this is divisive. For Giddens, leaving people reliant on benefits isolates them from broader society and removing benefits to force people into work (conditionality) forces them into an already overcrowded low-skilled, low-paid labour market; exploited by the systems set up to help them; trapped on benefits and let down by a failing education system and National Health Service (le Grand, 1999). In this regard, Third Way policy has a strong moral vein, the individual has a responsibility not only to themselves but also to the community. For example, jobseekers were required to show they were looking for work in exchange for receiving financial support from the state, in exchange for improved education and schools, parents have a responsibility to ensure homework is being done. The Third Way sees the relationship between the individual and the state as one of reciprocity, the state will support you in your ambitions; but the onus is on you to work hard and earn reward.

New Labour and housing policy

The election of New Labour (under Tony Blair) in 1997 saw a change in the fortunes of social housing. Although New Labour continued the policy of seeing home ownership at the top of a 'housing ladder', they also initiated a major investment in the social housing stock under the Decent Homes Programme (DETR, 2000), invested in homelessness prevention and housing-related support via the Supporting People programme, making significant progress on tackling homelessness and rough sleeping. The Thatcherite 'right to buy' policy continued as did LSVTs) and creation of ALMOs to transfer housing stock out of direct local authority control. These measures further residualised social housing as a tenure for those on the lowest incomes (CIH, 2018).

Malpass and Victory (2010) argue that the trajectory of welfare reform continued under the New Labour governments, although there was recognition that welfare and housing policies needed to be better aligned. Pawson and Mullins (2010) argue that New Labour perceived council housing as a redundant project with low demand and poor-quality stock and saw the issue through the lens of poverty and limited life chances and opportunities, not merely as a matter of giving tenants a choice of landlord. Labour recognised there was a stigma attached to the widening difference between homeowners and those living in estates with low demand and poverty, with the term 'sink' estate used by many commentators. Tony Blair stated in "Bringing Britain Together – a national strategy for neighbourhood renewal",

> Over the last two decades the gap between these worst estates and the rest of the country has grown…It shames us as a nation, it wastes lives and we all have to pay the costs of dependency and social division.
>
> *(Cm 4045, 1998, p. 1)*

The stigmatising of social housing tenants and the individual impact was outlined by Hanley in her experience of growing up in social housing and the impact it had on her life, "Estates is a bruise in the form of a word: it hits the nerves that register shame, disgust, fear and, very occasionally, fierce pride" (Hanley, 2012, p. 20). She goes on to summarise the feeling of social exclusion in social housing for many,

> Estates mean alcoholism, drug addiction, relentless petty stupidity, a kind of stir craziness induced by chronic poverty and the human mind caged by the rigid bars of class and learned incuriosity.
>
> *(Hanley 2012, p. 7)*

In conjunction with stigma of social housing put forward by Hanley there is also the juxtaposition of its scarcity.

> The truth is that council housing is a living tomb. You dare not give up the house because you might never get another but staying is to be trapped in a ghetto of both place and mind.
>
> *(Hutton 2007)*

New Labour did, however, recognise the need to improve the housing product on offer. The Housing Green Paper, "Quality and Choice: A Decent Home for All" (DETR, 2000) aimed to bring all social housing into decent condition by 2010. Local authorities with housing stock were required to undertake an options appraisal to identify how they would secure resources to meet the new

Decent Homes Standard. For many, the only option in raising funds to improve homes was to consider an LSVT. This continued the Conservative approach of removing housing from the public sector with almost a million homes removed from local authority ownership via LSVT between 1997 and 2009. New Labour reflected the Conservative approach in that it saw the state in an enabling role in relation to housing as opposed to necessarily providing and managing it.

Coalition government

If the transition from the Conservatives to New Labour in 1997 was a largely uneventful coup (if it was a coup at all) (see King, 2006a), the switch back to a more neoliberal policy approach under a Conservative and Liberal Democrat coalition, was anything but. The financial crisis of 2008, the roots of which had been in train for some time (Blonde, 2010), exposed banks, deemed 'too big to fail' as having allowed themselves to get into a situation where they did not have enough reserve capital to meet their debts. Governments across Europe were forced to bail them out. This recapitalisation alongside the economic downturn meant that the UK government (and others) had to take on billions of pounds in debt from the international markets. New Labour had already incurred significant public debt because of high levels of public spending. In 1997, Labour inherited national debt of under 40% of GDP. By 2003, they were running a budget surplus, adhering to Conservative spending plans (as they had promised to, to fend off accusations of profligacy). New Labour began to run increasing budgetary deficits between 2003 and 2008, during the biggest boom in GDP on record, despite Gordon Brown's claims of prudence.

This expansion of government debt led to the incoming Conservative and Liberal Democrat coalition government implementing considerable cuts to government spending to reduce the budget deficit (the annual increase to the national debt), and eventually the national debt itself. Even the former Labour Chancellor of the Exchequer, Alistair Darling, was warning there would have to be cuts to public spending worse than those made under Thatcher (Darling, 2012). Under the rubric of reducing public spending (or austerity), the coalition government embarked upon an ambitious programme of public service reform – neo-liberalism on steroids. Indeed, governments across Europe retrenched public spending under the banner of 'austerity', calling for government and individuals to live within their means (Farnsworth & Irving, 2015).

The coalition and housing policy

In terms of housing policy, the coalition government continued with the previous Conservative (and arguably New Labour) predilection for home ownership. The coalition government failed to govern the housing system, having given up some of the powers to do so via the localisation of the planning system and budget cuts (Tunstall, 2015). Further, the coalition continued the trend of shifting expenditure from object (houses) to subject (people) subsidies, this did not deliver anywhere near the levels of affordable (social) housing needed (Tunstall, 2015).

Reduced government support and increased self-reliance drove housing and welfare policy. Increased conditionality in the system included the use of sanctions (non-payment of benefits) for those who did not meet the requirements set out for them in their agreement with DWP. In an echo of Octavia Hill's housing system, the introduction of a more punitive system was intended to push people towards behaving in the way the state favoured. The coalition made significant and lasting changes to the welfare system that had profound effects on housing policy (if housing policy was distinguishable from welfare policy by this point).

Housing Benefit was folded into Universal Credit (UC) alongside five other means-tested benefits and tax credits as a single monthly payment. Local Housing Allowance (LHA) (Housing Benefit for the private rented sector) was frozen (CIH, 2018). The 'removal of the spare room subsidy' or 'bedroom tax,' as it has become known, was an attempt to deal with the perceived problem of households with more rooms in their homes than they needed, a disparity between social and private tenants. Social tenants could have the full rent paid on their homes regardless of whether they needed a home of a given size, private tenants were given an allowance according to their household size and had to make up any shortfall if they wished to rent a property larger than their assessed need. People over the eligible age for pension credit were, however, exempted. An overall benefit cap was also introduced. Originally set at £26,000, the level of median household earnings reflected the government's conviction that no household on benefits should receive more than this, the average household income, on equity grounds. This has since been decreased to £20,000 outside and £23,000 inside London – these are completely arbitrary figures.

In terms of outcomes, there is ample evidence that these measures did not even meet their own policy objectives. UC was found to diminish work incentives (Commonweal Housing, 2017), meeting the conditions set within UC proved problematic and a bureaucratic nightmare for those working multiple jobs, irregular hours etc. – leading to further sanctions (Commonweal Housing, 2017; Wright et al., 2018). Sanctions did little to motivate homeless people to seek or prepare for work, caused considerable distress and might have pushed some extremely vulnerable people out of the social safety net altogether (Johnsen et al., 2018). Alongside other welfare reforms, UC made many households worse off than under the system it replaced (Schmuecker, 2017). The impact of the bedroom tax has been harmful and has not addressed its own policy goals either (Gibb, 2015). Households have less space than they often need; this particularly affects lone parents with child access arrangements and people with disabilities (Wilcox et al., 2017). The Benefit Cap piled on further misery. Research by the Chartered Institute Housing (CIH, 2017) found that the effect of the cap was disproportionately felt by larger families and was causing real hardship and in the worst cases homelessness. Children have been disproportionately affected despite having no control over their circumstances. The cap has limited the ability of larger families to meet rents in some areas. Fitzpatrick et al. (2018) found that four out of five households with four children faced at least a £50 shortfall in their rent, rising to over 90% for five child households (CIH, 2016). CIH also found that the cap is not making work the best route out of poverty (as promised), is manifestly unfair in some cases, e.g. where children have been fostered and impacts heavily upon households with no means of escape; for example, where there are very young children (CIH, 2017).

Levelling up?

Following a general election in late 2019, the Conservatives were returned to the government with a parliamentary majority of some 80 seats. The manifesto made several commitments in this area (CIH, 2019; Conservative Party, 2019). First, to bring forward a white paper that sets out measures to empower tenants and support the continued supply of social homes and second to build one million homes over the five years of the parliament. There are also commitments to support home ownership by introducing long-term, fixed-rate mortgages, deposits of only 5% of a house's value, and expanding the 'First Home Programme' (CIH, 2019, p. 2) for local buyers to access a 30% discount. The Right to Buy scheme for Councils is to be continued and joined by a similar scheme for Housing Association homes (following a successful pilot [Conservative Party, 2019]). Private renting will also be reformed by enabling lifetime rental deposits, ending section 21 (no fault) evictions, and making changes to the balance of power between tenants and landlords in favour of the former (where they adhere to the terms of the tenancy). At the time of writing, the publication of the Social Housing Regulation Bill is imminent, which should action many of these initiatives.

The Social Housing White Paper

In November 2020, the government did publish the promised white paper entitled 'The Charter for Social Housing Residents' (MHCLG, 2020). Alongside measures to improve building safety following the Grenfell Tower fire, the white paper commits a number of substantive actions.

Building safety

The Building Safety Bill is currently making its way through Parliament, where amendments in the House of Lords extend government powers to pursue developers for remediation costs and limit the liability of leaseholders. Relatedly, the Health and Safety Executive (HSE) continue to set up the Building Safety Regulator (BSR) to oversee building safety in high-rise buildings. Changes have also been made, giving building owners more flexibility in how they assess and manage fire safety risks. To help promote fire safety, the government is also bringing forward regulations to require smoke and carbon monoxide detectors in all rented homes.

Regulation

A new tripartite regulatory arrangement for social housing continues to take shape.

- The Housing Ombudsman Service, largely operating under its new remit, is publishing a wide variety of reports, data, and training materials
- The Regulator for Social Housing (RSH) has published the principles it will apply in regulating the sector, and the draft outcomes it will be assessing for. Once the government passes the 'Social Housing Regulation Act', giving shape to the role of the regulator, there will be further consultations on the regulatory standards and how the inspection regime will operate. RSH is already consulting on the proposed 'Tenant Satisfaction Measures' that will be used to assess landlords
- The Future Homes Standard continues to shape environmental standards for new homes, with the building regulations being revised to support its goals.
- An updated Decent Homes Standard, proposals for electrical safety checks in social housing, vehicle charging points in new homes and greater protections for private renters are also forthcoming. However, planning reform continues to be delayed after the proposals promised in the 'Levelling Up' White Paper, were not included.

Professionalisation and stigma

Embedded throughout the Social Housing White Paper, professionalisation and stigma overarch many of the other themes. The government has convened a working group to review the training and staff development of social housing staff. The Chartered Institute of Housing have launched a professional standards framework for people working in housing, and it seems likely that this may form a key part of any new approach to professional development coming out of the working group.

Tenant engagement and voice

The RSH is building on the National Housing Federation's (NHF) 'Together with Tenants' report to engage with tenants about its activities. Once the Social Housing Regulation Bill is passed, there will be more clarity about expectations from landlords, but the report lays out the probable areas that will be covered. Like professionalisation and stigma, this is an overarching theme, running throughout building safety, regulation, and zero-carbon measures.

Funding

There is something of a puzzle around landlords being able to fund themselves adequately to carry out their core functions alongside the myriad other expectations of the government, their tenants, and others. Building safety remediation, maintaining ageing stock and zero-carbon all place significant financial costs on landlords, whilst their ability to raise income is limited by the rent formula and other factors. Social landlords also have limited recourse to grant remediation funding, that others do have access to. In addition, raising finance from both the state and private markets will increasingly become reliant on meeting Environmental, Social and Governance (ESG) targets, with interest rates used to incentivise compliance

New supply

The House of Lords Built Environment Committee released a report on meeting housing demand. The report's recommendations span assessing and calculating need, getting the balance of tenure right, training enough qualified staff, and the quality and design of new homes. The current 'Affordable Homes Programme' runs until 2026, making this a key time to begin influencing what replaces it.

'Everyone In'

On homelessness specifically, the government's response to rough sleeping during the COVID-19 pandemic, the 'Everyone In' scheme is credited with saving lives and virtually ending rough sleeping in England over a single weekend. The Government committed an initial £3.2m for this, and then a further £105m to provide temporary accommodation for people at risk of rough sleeping during the pandemic. Crucially, Government also lifted 'no recourse to public funds' rules meaning that people not eligible for support because of immigration rules were able to access accommodation and support. Other measures such as a temporary increase to UC, returning the LHA rate to the 30th percentile and stay on court actions were also crucial in stemming the follow of potential homeless households.

To begin to learn lessons from the pandemic response, the government set up a commission chaired by former head of the Home Civil Service, Sir Bob Kerslake. The commission received submissions from a range of organisations and individuals and carried out a series of bi-lateral meetings to gather evidence. The headline recommendations of the report (Kerslake Commission, 2021) cover central government taking a more strategic approach to ending rough sleeping, placing specific focus on prevention instead of crisis management. Kerslake also recommended maintaining changes to welfare benefits and ensuring that limits on public funds for non-UK nationals stop short of destitution. The commission recommends that local authorities produce long-term, integrated homelessness and health plans alongside local homelessness partners agencies, with the aim of quantifying the level of government funding required. He also recommended that Homeless Link consult on a professional accreditation framework for staff working with people experiencing homelessness, that recognises the complexity of the sector and skills needed. The report recognises that housing providers are not considered 'public bodies' and are therefore under no legal duty to address homelessness. However, the report recognises the social and moral duty that housing providers have.

So why is there homelessness and poor-quality housing, in England in 2022?

In the first part of this chapter, I have tried to set out some of the main theoretical approaches to social justice that have guided policies and decisions about the distribution of resources, including housing. These operate on something of a spectrum. Looking back to the opening chapter, socialism and communism offer a conception of social justice where private property rights are severely curtailed or even abolished altogether. Whilst libertarian ideas, such as those of Robert Nozick (1974) view property rights as sacrosanct, where exchange has been fair, outcomes are irrelevant to the justness of who owns resources. In the middle, we have Rawls, who attempts to balance the wellbeing of the poorest with the rights of all to own resources commensurate with their skills, talents and abilities.

In the second part of the chapter, I have attempted to give a somewhat potted history of housing provision in the UK and latterly England, that shows how these ideas have influenced the policies of various governments in relation to housing provision, and indeed the distribution of resources more generally.

In short, the answer to the question is that different political parties have had very different ideas about justice, social justice and how resources are distributed. During the feudal period, thinkers such as Filmer and Aquinas justified the large disparities in wealth between royalty, the aristocracy and 'serfs' as the will of God. The people were seen as the possession of the monarch, with the biological analogy of the body used to explain the role of the family structure. Most were left to earn what they needed to live on, with little to nothing available from the state to support them. Charity, of course, was one way that the rich could get closed to God and donations to the poor were often seen as an obligation under the doctrine of *noblesse oblige*. In the later feudal period, the Poor Laws began to make provision for the poor, but this was far from what we would consider fair or adequate in today's terms.

Following the *Glorious Revolution* and institution of (largely) democratic government, there was gradual change in provision for the poorest people – which was most of them. Another revolution, the Industrial Revolution, kickstarted greater provision for the poor. Although, it mostly took the form of punitive criminal punishments for vagrancy, the workhouse, or asylums for those who disrupted the workplace; and latterly anyone who was perceived, in some way, to be deviant. Provision of individual homes only came onto the agenda towards the end of the 19th century, its role boosted by soldiers returning from the Great War and the Russian Revolution, both homes for heroes, and to stave off actual revolution. It could be said that this was driven by both notions of social justice for returning soldiers, and a degree of self-preservation by the upper classes.

Housing as a serious (wobbly) pillar of social justice was only realised following World War II. Soldiers returning from the war, alongside a genuine concern for social justice, drove the inception of the welfare state, and a massive house building programme, mainly via the state. Although there was a genuine determination that there should be a standard below which no person should fall, council housing between 1945 and the 1970s was for those who could afford to pay the rent, the affluent working class, with local authorities able to give limited rebates. There was also no security of tenure for council tenants. We now consider this normative, and a right. Post-war housing provision then was sometimes not the egalitarian programme it is sometimes painted as.

The election of Margaret Thatcher in 1979 took housing policy, and government policy more generally, back to far more individualistic place. Council housing was viewed as an unnecessary state intrusion on individual liberty and the government embarked on a mass selloff of council stock, that we are still feeling the impact of today. Between 1979 and 1997, Conservative governments generally implemented policies that looked to the individual to take responsibility to provide for themselves and the acquisition of wealth was viewed in far more Nozickean terms than previously. Many restrictions on the labour force were removed, the power of the trade unions reduced and state-owned services were privatised via the stock market (Steger, 2003). Lockean notions of negative rights and freedoms were prized over Rawlsian or Senninan notions of social justice. For example, rough sleeping (especially in London) spiked with 'cardboard cities' of rough sleepers seemingly ubiquitous. The government view was that this was a result of some personal failing of the individual, rather than social failure.

(New) Labour returned to the government in 1997 with a new 'Third Way' conception of social justice. Rights now came with commensurate responsibilities; this is a fundamental change to the notion of 'rights' which had previously largely been thought of as unconditional. Also, there could be no authority without democracy – traditional sources of authority were therefore denigrated. This is also a fundamental break with particularly the conservatism of Oakeshott (1962) and Scruton (2001; 2006) that recognised the ordering force of religion and other social structures. It was also a break with 'old' left thinking that had prized the role of civil society organisations such as trade unions and mutual societies. Although they largely continued with Conservative housing policy, e.g. Right to Buy, they did at least recognise that the social housing stock needed updating and embarked upon the Supporting People programme to support people to get into and maintain independent living. However, the latter certainly embodied the 'no rights without responsibilities' framework with its plethora of contracts, agreements and support plans (see Seddon 2008; Irving-Clarke, 2019).

The return of the Conservative Party, first in coalition with the Liberal Democrats and then as the sole party of government, saw a return to a more individualistic conception of social justice and a more punitive social security system that attempted to coerce people into 'correct' behaviours via the use of wider and more severe conditionality in the welfare system. Rhetoric around 'shirkers and strivers' was commonplace. Having shrunk to all-time lows under New Labour, homelessness once again began to increase as local authorities felt the impact of cuts to public spending, the welfare state also became markedly less generous. However, the impact of the COVID-19 pandemic made homelessness explicitly a public health issue, as the Conservative government funded local authorities to get rough sleepers off the streets and into self-contained accommodation. Limits on recourse to public funds were also lifted, meaning that all rough sleepers could be housed using these funds. Rough sleeping was effectively ended over a single weekend. The Conservatives also seem to have discovered some zeal for public spending with their 'levelling up' agenda and programme. Significant funding and powers are to be devolved to regional and local governments.

In conclusion then, homelessness and rough sleeping are largely a function of ideas about social justice and more pointedly over who is responsible for dealing with the problem. More socially minded and communitarian governments put more priority, time, and resources into resolving homelessness as an issue of social justice, whilst more individualistic governments view the issue through a lens of personal responsibility and self-discipline, more a view of 'justice' in its original definition of getting what one merits or deserves. However, more recently the pandemic the government was forced to take a public health approach to the issue, a break from any notion about justice at all. This led to the ending of rough sleeping almost overnight, and perhaps offers a different perspective through which to view the problem.

Suggested further reading

Rawls, J (1971) *A Theory of Justice*, Harvard University Press, Cambridge (Mass).
Taylor, H (2018) *Social Justice in Contemporary Housing*, Routledge, Abingdon.
Watts, B & Fitpatrick, S (2017) *Welfare Conditionality*, Routledge, Abingdon.

References

Alkire, S (2002) *Valuing Freedoms: Sen's Capability Approach and Poverty Reduction*, Oxford University Press, New York.
Aquinas, T (2014) *Summa Theologica*, Catholic Way Publishing, Online.
Beveridge Report (1942) *Social Insurance and Allied Services*, HM Government, London.
Blonde, P (2010) *Red Tory*, Faber, London.
Busch-Geertseema, V & Sahlin, I (2007) *The Role of Hostels and Temporary Accommodation*, FEANTSA, Brussels.
Central Housing Advisory Committee (CHAC) (1953) *Transfers, Exchanges and Rents*, HMSO, London.
CIH (2016) *Mind the Gap*, CIH, Coventry.
CIH (2017) *Feeling the Pinch*, CIH, Coventry.
CIH (2018) *Rethinking Social Housing*, CIH, Coventry.
CIH (2019) *What You Need to Know About Housing and General Election 2019*, CIH, Coventry.
Clarke, J et al. (eds) (2000) *New Managerialism, New Welfare?*, SAGE Publications, London.

CM 4045 (1998) *Bringing People Together: A National Strategy for Neighbourhood Renewal*, HMSO, London.

Commonweal Housing (2017) *Commonweal Annual Impact Review 2016/17*, Commonweal Housing, Elmbridge.

Conservative Party (1951) *General Election Manifesto*, Conservative Party, London.

Conservative Party (2019) *Get Brexit Done: Unleash Britain's Potential. The Conservative and Unionist Party Manifesto 2019*, The Conservative Party, London.

Cowan, D & Marsh A (2018) A perennial problem? On underoccupation in English Council Housing, *Housing Studies* 34(2) pp.319–337.

Crocker, D.A. (2008) *Ethics of Global Development: Agency, Capability and Deliberative Democracy*, Cambridge University Press, Cambridge.

Crowther, A (1981) *The Workhouse System 1834–1929: The History of an English Social Institution*. Batsford Academic and Educational, London.

Cullingworth Report (1969) *Cullingworth Committee Report on Council Housing: Purposes, Procedures and Priorities*, HMSO, London.

Darling, A (2012) *Back from the Brink*, Atlantic Books, London.

DETR (2000) *Quality and Choice: A Decent Home for All*, HMSO, London.

Driver, S & Martell, L (1999) Left, right and third way, *Policy Studies* 28(2) pp.147–161.

Farnsworth, K & Irving, Z (2015) *Social Policy in Times of Austerity*, Policy Press, Bristol.

Filmer, R (2016) *Patriarcha*, Createspace, Scott's Valley CA.

Fitzpatrick et al. (2018) *The Homelessness Monitor: England 2018*, Crisis, London.

Fitzpatrick, S and Watts, B (2010) The Right to Housing for Homeless People, *Homelessness Research in Europe*, FEANTSA, Brussels.

Fitzpatrick, S, Bengtsson, B and Watts, B (2014) Rights to housing: Reviewing the terrain and exploring a way forward, *Housing, Theory and Society* 31(4) pp.447–463.

Fitzpatrick, S and Watts, B (2018) *Welfare Conditionality*, Routledge, Abingdon.

Foye, C (2017) The relationship between size of living space and subjective well-being, *Journal of Happiness Studies* 18 pp.427–461.

Gibb, K (2015) The multiple policy failures of the UK bedroom tax, *International Journal of Housing Policy* 15(2) pp.148–166.

Giddens, A (1998) *The Third Way: The Renewal of Social Democracy*, Polity Books, Cambridge.

Gough, I (1980) Thatcherism and the welfare state: Britain is experiencing the most far-reaching experiment in 'new right' politics in the western world in *Marxism Today* pp. 7–12.

Gurney, C (1999) Pride and prejudice: Discourses of normalisation in public and private accounts of home ownership, *Housing Studies* 14 pp. 163–185.

Hanley, L (2012) *Estates: An Intimate History*, Granta Books, London.

Henderson K (2019) *The Role of Housing in a Coordinated Community Response to Domestic Abuse*, Unpublished PhD Thesis, University of Durham.

Hayek, F (1982) *Law, Legislation and Liberty*, Routledge and Kegan Paul, London.

Hutton, W (2007) *Open the Gates and Free People from Britain's Ghetto, The Observer*, February 18, 2007.

Irving-Clarke (2016) *Supporting People: How Did We Get Here and What Does it Mean for the Future?* Unpublished PhD Thesis, De Montfort University, Leicester.

Irving-Clarke (2019) *Supported Housing: Past Present and Future*, Routledge, Abingdon.

Johnsen, S et al. (2018) *Conditionality Briefing: Homelessness and 'Street Culture'*, University of York, York.

Kant, I (1997) *A Critique of Pure Reason*, Cambridge University Press, Cambridge.

Kerslake Commission (2021) *When We Work Together: Learning the Lessons*, Kerslake Commission, London.

King, P (2003) Housing as a freedom right, *Housing Studies* 18(5) pp.661–672.

King, P (2006) *Choice and End of Social Housing*, IEA, London

King, P (2006a) *A Conservative Consensus: Housing Policy Before 1997 and After*, Imprint Academic, Exeter.

King, P (2010) Housing as a freedom right, *Housing Studies* 18(5) pp.661–672.

King, P (2011) Using big ideas: The application of political philosophy in housing research, *Housing Theory and Society* 28(2) pp.109–122.

King P, (2012) Social justice, in S.J. Smith, M Elsinga, L.F. O'Mahony, O.S. Eng, S. Wachter, & C. Hamnett (eds), *International Encyclopaedia of Housing and Home, Vol 6.* Elsevier, Oxford, pp.464–468.

le Grand, J (1999) *CORA and the Third Way: Some Unanswered Questions*, LSE, London. Locke, J (2016) *Two Treatise of Government*, Createspace, Scott's Valley CA.

Malpass, P & Murie, A (1982) *Housing Policy and Practice*, MacMillan, London.

Malpass, P & Victory, C (2010) The modernisation of social housing in England, *International Journal of Housing Policy* 10(1) pp.3–18.

Mead, L (1992) The New Politics of Poverty: The non-working poor in America, *City Journal* Summer 1992.

MHCLG (2020) *The Charter for Social Housing Residents: Social Housing White Paper*, HM Government, London.

Mill, J.S. & Bentham, J (1987) *Utilitarianism and Other Essays*, Penguin Classics, Hammondsworth.

Miller, D (1976) *Social Justice*, Oxford University Press, Oxford.

Mullins, D & Murie, A (2006) *Housing Policy in the UK*, Palgrave, London.

Neale, J (1997) Homelessness and theory reconsidered, *Housing Studies* 12(1) pp.47–61.

Newman, J.E. (2001) *Modernising Governance: New Labour, Policy and Society*, Sage, London.

Newman, J.E. (2014) Landscapes of antagonism: local governance, neoliberalism and austerity, *Urban Studies* 51(15) pp.3290–3305.

Nozick, R (1974) *Anarchy, State and Utopia*, Blackwell, Oxford.

Nussbaum, M (2011). *Creating Capabilities: The Human Development Approach*, Harvard University Press Cambridge, MA, pp.33–34.

Oakeshott, M (1962) *Rationalism in Politics and other essays*, Methuen, London.

Pannell, J & Thomas, C (2000) *Almshouses into the Next Millennium: Paternalism, Partnership, Progress?* Policy Press, Bristol.

Park, J (2017) *One Hundred Years of Housing Space Standards*, Online at: http://housing-spacestandards.co.uk/assets/space-standards_onscreen.pdf. (Accessed 29 March 2022).

Pawson, H & Mullins, D (2010) *After Council Housing: Britain's New Social Landlords*, Palgrave MacMillan, London.

Plato (1971) *The Republic*, Penguin Classics, Hammondsworth.

Rawls, J (1971) *A Theory of Justice*, Harvard University Press, Cambridge, MA.

Schmuecker, K (2017) *Universal Credit: A Joseph Rowntree Foundation Briefing*, Joseph Rowntree Foundation, York.

Scruton, R (2001) *The Meaning of Conservatism*, Palgrave, Basingstoke.

Scruton, R (2006) *Arguments for Conservatism: A Political a Philosophy*, Continuum, London.

Seddon, J (2008) *Systems Thinking in the Public Sector: The Failure of the Reform Regime and a Manifesto for a Better Way*, Triarchy Press, Bridport.

Sen, A (1974) Informational bases of alternative welfare approaches: Aggregation and income distribution, *Journal of Public Economics*, 3(4) pp.387–403.

Sen, A (1979) Equality of what? in McMurrin (ed), *Tanner Lectures on Human Values*, Cambridge University Press, Cambridge, pp.197–220.

Steger, M.B. (2003) *Globalization: A Very Short Introduction*, Oxford University Press, Oxford.

Stephens, M et al. (2018) *UK Housing Review*, CIH, Coventry.

Taylor, H (2018) *Social Justice in Contemporary Housing*, Routledge, Abingdon.

Torgerson, U (1987) Housing: The wobbly pillar under the welfare state, *Scandinavian Housing and Planning Research* 4 pp.116–126.

Tucker (1966) *Honourable Estates*, Gollancz, London.

Tunstall, R (2015) *The Coalition's Record on Housing Policy, Spending and Outcomes 2010–2015*, Centre for Housing Policy, York.

Warburton, N (2014) *Philosophy: The Classics*, Routledge, Abingdon.

Watts, B & Fitzpatrick, S (2018) *Welfare Conditionality*, Routledge, Abingdon.

Wilcox, S et al. (2017) *UK Housing Review*, CIH. Coventry.

Wolff, J (1991) *Property, Justice and the Minimal State*, Stanford University Press, Stanford, CA.

Wolff, J & de-Shalit, A (2007) *Disadvantage*, Oxford University Press, Oxford.

Wright, L et al. (2018) *Cost Effectiveness of Housing First*, What Works Centre, HMSO, London.

Yorke, T (2017) *Homes Fit for Heroes: The Aftermath of the First World War 1918–1939*, Countryside Books, Newbury.

5

WHY CAN'T I BURN MY HOUSE DOWN IF I FEEL LIKE IT?

> **Case study**
>
> I want to burn down my house. I don't like living there anymore. No one else lives with me and the house is solely mine.
>
> So, why can't I just burn my house down if I feel like it?
>
> - Are there any reasons the law should stop me from doing so?
> - Are there any moral reasons I should not do this?
> - What externalities might this cause that mean I should not do this?

Introduction

If property rights mean anything, it is surely the right to control and dispose of the items that we own, in a manner that we choose.

If we think about the smaller things that we own, clothes, books, ornaments etc. we can see that, generally, these are ours to dispose of as we wish. Indeed, one of common criticisms of modern life is that many things are too disposable. Take clothing for example, cheap mass market retailers such as Primark offer relatively cheap fashion and turn over their stock quickly. The quality probably isn't the highest and if you buy something, wear it a few times and it tears, or gets damaged in the washing machine, it isn't the end of the world, and you haven't wasted too much money. Alternatively, if you spend more money on an item of clothing, it should last a lot longer and is probably better value for money. But, nevertheless, these items are yours to dispose of as you choose to. You can throw them away, donate them to charity, or burn them, as you wish.

DOI: 10.1201/9781003298212-6

However, there are some larger more valuable items, a car, or a house for example, that are not so disposable. First, they are physically larger and so harder to dispose of, they are also considerably more valuable and so you do not want to dispose of them, even when they fail to function correctly. Further, they are worth repairing, they are probably even worth paying someone else with greater expertise to repair for you. For example, if your car breaks down, you probably aren't going to just scrap it. You might if it's an old banger and not worth repairing, but in general, you'd take it to a garage and get it fixed. There are even people who persist in fixing cars when they are long past their useful life, this might be because they enjoy the process, or it might be that it was their first car or has some other sentimental value. Humans are not always rational creatures after all.

You are probably also unlikely to discard your house if the heating breaks, or some tiles fall off the roof, or plaster falls of a wall. This is most likely because your house is the most expensive and financially valuable thing you own. It would be a terrible waste of money to do so. Your house is probably also your primary asset, and where most of your wealth is stored.

This chapter explores the origins of property rights, what they mean and the limits of them. It looks at the ideologies explored in Chapter 1 and at the different views on ownership of property found within those ideological frameworks. I will also touch on the role of the state from Chapter 2, and the state's role in granting and enforcing property rights. What are property rights? How did the right to own property come about? Do we all agree that the right and ability to own property are a good thing? This chapter attempts to answer these questions, and others, using both theory and practical examples. Before turning to the titular question, can I just burn my house down, if I feel like it?

What are 'property rights'?

What does it mean to 'own' something? How is this different from (in the case of housing) renting? When a person or household is granted a tenancy, they are given sole use of that property, just like an owner, this may be for a short or extended time, but nevertheless there is a qualitative difference in 'ownership, versus 'renting'. Is there, by association then, qualitative difference in how owners and tenants feel about their homes?

Before we begin to discuss property rights, we should spare some time to consider what rights are more generally more fully than we have thus far. Previously, I have differentiated between 'negative' and 'positive' rights. Negative rights being the more usual view of rights in that all that is required is for other not to prevent us from exercising our rights. The right to life, freedom from torture, and indeed the right to own property are all examples of negative rights. 'Positive' rights on the other hand are rights which require action from others to enact. For example, the right to healthcare requires some entity to provide it, to train and employ Drs and nurses to administer it. Likewise, a right to housing would require some entity to build and distribute it.

More generally, rights are usually stated as a formula 'A has a right to do B', what this suggests, as above, is that others have a duty not to prevent A from doing B, and that A doing B, protects some interest of A's (King, 2003). This brings in the idea that existence and social recognition of 'rights' confers 'duties' on the rest of us not to impinge upon those rights. It also means that A, in protecting their self-interest, should feel no embarrassment about enforcing their claim and the duties others owe them. Thus, a 'right' is a legitimate claim one can enforce against others, including organisations and the state (Waldron, 1993). This shows that rights, far from being abstractions, are the basis of social relations (Machan, 1989). This also shows, importantly, that one does not merely have 'rights' but the 'right to' something significant or important (Freeden, 1991).

Thus, rights depend on three pre-existing conditions. First, that there are good reasons for demanding the 'thing', second, there is something to enjoy, be it life, liberty or even housing; and third, that certain social arrangement must be made to enforce our right to that 'something' (King, 2003). Further, talking about rights is another way of talking about the responsibilities of others to us. This is particularly clear in Nozick's (1974) discussion of 'side constraints' where our duties to each other constrain our actions. Rights are also discussed in universal terms, that is to say that we all have a right to something, this means that we are all subject to the same 'side constraints' on our behaviour. Rights then, cannot be limitless, we have to moderate our behaviour to not infringe on the rights of others (King, 2003). Lastly, because rights are universal, we should be concerned with the rights of others, as if they were our own; because one day, they might be.

The institution of 'property rights' can be said to entail three things (King, 2016). First, when we own something, it is at our disposal, and we may use it or dispose of it as we wish to. This does not of course mean we can do what we like with it. As King (2016) points out, we cannot own a knife and stab someone with it, as this is a violation of their right to life. We might, of course, do so if that person is threatening our 'right to life', or that of another person; in such a circumstance, stabbing them can be seen as a legitimate act. Although, if they are stealing something that belongs to us, and are not posing a threat, stabbing them may well be seen as illegal and punishable. For example, if someone steals your bike and is riding them away, shooting them in the back is unlikely to be seen as a proportionate response. The right own 'chattels' then is not absolute in the sense that deadly force will probably not be seen as justifiable in retaining our chattels.

Second, if I 'own' something, it is for my exclusive use. I can decide who may or may not use that thing and can quite justifiably control who has access to that thing. My right to do this, therefore, must be backed by some institution that protects my rights and punishes transgressors (King, 2016).

Third, and this is what separates owning from renting, I can dispose of an item as I wish to. I may give it away, sell it, transfer it to another person as I see fit. Renters, despite being given sole access or use of a rented item, are not legally entitled to do this. There are some limits on whether or how I may dispose of my property. I am generally free to sell anything I own, but I (rightly) cannot refuse to sell to someone because they are black or Jewish for example (King, 2016).

The enforcement of property rights, of course, relies on the existence of some institution able to recognise property rights and enforce them; there also has to general recognition in society that property rights exist, and recognition of proof of them (King, 2016). The most obvious example of this is the deeds of ownership you might have to your house, or a tenancy agreement to prove you have the right to occupy. Likewise, in the UK, if you own a car, the V5 document proves your ownership of the vehicle. If you purchase something from a shop, you are provided with a receipt for it to prove your legitimate ownership. Of course, these documents are not only to prove ownership, but also to ascertain responsibility should your car or some other possession be used to break the law. In some cases, e.g. traffic offenses, you can be required to tell the authorities who used your vehicle, if, indeed, you know. If you wish to sell your car or house, you will be required to produce the V5 certificate or deeds to prove your right to do so, via ownership. King (2016) makes a comparison with former communist nations turning to Hernando De Soto (2000) to show how the lack of such a process in these nations prevents the generation of wealth. De Soto shows that people and businesses in some former communist countries in Latin America lack the ability to prove ownership and therefore use their property as leverage to invest in their businesses. Poorer people in these societies, according to De Soto, live in houses that are on land where ownership cannot be determined. They are therefore unable to use the value of their home to claim their wealth. On the other hand, and as De Soto points out, the prime source of entrepreneurial funding in the US is a mortgage on the entrepreneurs' home, thus providing funding for their business and hopefully an increase in their overall wealth. This also points to the fact that housing is a store of wealth, it is quite fashionable to say that housing should not be used as such, but this ignores the long history of this being the case and the fact that this is an obvious use of housing and long has been. For a start, no one has much of a problem with social landlords leveraging the value of their stock to build new social housing, so why should private owners be so censured? There are, of course, some serious points to be about the impact of second homes, or the use of very valuable properties as stores of wealth gained nefariously, but this should not lead us to denigrate the use of housing as a store of wealth per se.

Origins of property rights

Property rights, the right to own private property, are a given across most of the world in 2022, as this is being written; and likely to remain so. But this was not always the case and the right to own property has been hard fought for over a long period of time. If we think back to Chapter 1 and the discussion about feudal governance and the views of Thomas Aquinas (2014), most people owned very little private property beyond personal effects. They might own their home, given that they probably built it themselves (more of which shortly), but this would have little, probably no monetary value. In any case, the monarch and likely the Lord of the land you lived on, likely had the right to confiscate anything you, as a serf, owned anyway. So, your right to own property was limited at best.

Before moving onto developments in practical property rights, we should spend some time looking at the theoretical background of these rights to understand how the right to private property can be justified. In any discussion of rights, we should pay attention to where these ideas come from. We should not, accept anything as 'self-evident' or as a 'given'. We might ask, for example, why is it self-evident that killing another person is wrong?

One response to this question might be that Genesis 1:1 26–27 states that all human beings are created in the image and likeness of God, and whilst the exact meaning of this is disputed, as most bible verses are, the essential principle is that humans represent God, and more importantly each person has a spark of the divine within them. It is therefore wrong for any one person to take the life of another. It might also be interpreted as God giving special status to humans, thus justifying their dominion over the earth and animals and plants. It's also true, of course, that in the ten commandments, it directly says 'Thou shalt not kill', albeit at only commandment number six. This is the root of property rights, the individual as a sovereign being with control over their own body and actions.

In Britain, property rights were first enumerated in 1215 with *Magna Carta*. This is widely considered one of the most important statements of rights in history. Magna Carta established the rights of all 'free citizens' to own and inherit property and to be protected from excessive taxes. It also established the right of property-owning widows to not have to remarry (hence passing their property to their new husband), and established principles of due process and equality before the law. These final two points are important to our discussion as due process and equality before law are important principles by which property rights can be defended.

Magna Carta was bolstered in 1628 by the *Petition of Right* as a statement of civil liberties. King Charles I's unpopular foreign policy had caused his government to exact forced loans and station troops in subjects' houses as an economy measure. Arbitrary arrest and imprisonment for opposing these policies led to violent hostility to Charles and to George Villiers, the Duke of Buckingham, in Parliament. The Petition of Right asserted four principles. First, no taxes may be levied without consent of Parliament, second, no subject may be imprisoned without cause shown (reaffirming the right of *habeas corpus* or due process), third, no soldiers may be quartered upon the citizenry, and lastly martial law may not be used in time of peace. These rights largely carry through to the present day.

Property rights are, of course, a function of living in a community. A person living on his own need has no concern about property rights as there is no one to threaten him nor prevent him from using anything he sees fit for his own benefit. However, as social animals, people have lived in community settings for millennia. Some rules over property are therefore needed. Moreover, there is a need for the sovereign to accept property rights and enforce these where they are legitimately claimed. If we think back to Thomas Hobbes and Leviathan (2014), the role of the sovereign is to prevent a 'state of war' between people by taking such steps as are necessary, with few limits, to safeguard the safety of people and protect property rights. There is an, at least theoretical, acceptance that without a sovereign to enforce them, rights mean very little. 'Nonsense on stilts' as Bentham (1987) might have it. Hobbes does not even particularly have much faith in people to enforce property rights, as far as he is concerned society would dissolve into a war of revenge without an all-powerful sovereign.

Locke (2016) takes a rather more optimistic view, seeing that man has a number of natural rights, owning property being one, that he will not transgress upon. Property rights are therefore, at least partly, protected by the virtue Locke sees as being inherent in man. Indeed, the US Constitution (1776) states that it can only be successful amongst a virtuous society. Locke also provides an explanation of how property rights can be practically defended, rather than being a mere theoretical construct endowed by an unseen creator. In founding the US, the founding fathers found themselves in a somewhat unique position in that, native American tribes aside, they found themselves in a largely unoccupied country where no person had jurisdiction or rights over the land, as they saw it. In the UK, we have never found ourselves in such a position, rather, the right to own property has had to be wrested from the Crown. Locke sets out how property rights can be practically acquired.

His thought was that natural materials, a piece of wood, or the land for example, could be 'mixed' with man's own labour, and as long as there was no competing claim, that property became his legitimately. For example, a piece of wood could be taken from the land and by carving, fashioned into a useful item, e.g. a bowl or walking stick. That item then became property by the mixing of the material and the labour required to fashion it. The same applied to land. If a piece of land was taken and transformed into a fertile growing area via irrigation and the working of it, that land could legitimately be the claim of the person carrying out the labour to farm it. Indeed, Locke further bolstered this idea with the thought that if by labour the land grew the same amount of crops as ten natural plots the same size, that conferred right of ownership via this productivity. Indeed, the history of the US is indeed one of frontiersmanship, with men and women gradually moving across the land making it productive. From this, trade of course follows, and we saw earlier this becomes the basis for the monetary and banking systems we see today. Indeed, the US has been transformed from a largely unoccupied, unproductive land to the one of the most prosperous countries ever to have existed, and in a relatively short space of time.

Whilst Locke sets out a process via which property can be acquired by labour, Robert Nozick (1974) is more concerned with transfer and exchange. However, his arch-libertarian theory certainly needs a basis from which private property can be legitimated. As we saw earlier, Nozick is largely concerned with legitimate property rights as the sole arbiter of justice. He is dismissive of the concept of social justice, insisting instead that justice is a function of legitimate exchange of property. If property is legally acquired, it cannot be interfered with by another party, even the state. This, of course, renders taxation illegitimate. Again, as we saw in Chapter 2, property rights are key to Nozick's (1974) acceptance of the need for a minimal state at all. He accepts that there is a need to protect property rights, but that this role is fulfilled by private companies contracting with individuals to protect their property. These companies compete for custom until only two remain, at some point, these two will combine into one organisation and thus the state is born, with a single body having responsibility for protecting everyone's property rights. For Nozick, one of the few roles of the state is to protect private property.

As I have already suggested, the recognition of property rights and the role of defending them is key to the whole concept. Property rights cannot exist without a protection from an independent arbiter. As we saw in the chapter on the State, a key part of the role of the police (in criminal cases) and the criminal and civil courts is to protect these rights, secure redress, and punishment when they are violated and arbitrate when there is a dispute. In the chapter on ideologies, we discussed different ideas of the state and of the role if private property. Here, we will look at the attitude of these ideologies to private property.

Conservatism and private property

As one might expect, conservatives and liberals consider the right to property as central to their thought. They generally consider that the state must be kept in check by institutions via a separation of powers, regular elections, and civil society organisations such as trade unions (Scruton, 2001). The state should generally not interfere in private property issues unless there is a dispute or crime that deprives an individual that they think is rightfully theirs. For conservatives, property takes on particular significance as it gives independence and dignity to the individual and prevents complete dependence on the state. In this manner, it acts as a 'check and balance' on state power over the individual.

Conservatives are inherently suspicious of the state and state intervention in people's lives, including state ownership of industries and homes. The right to own property is one of the bases of law and liberty, guaranteed by the laws and traditions of society. Conservatives also generally defend the ability of all to accumulate wealth and pass this on to their children when they die. It could be argued that conservatives view the acquisition of wealth as a proxy for a meritocratic society, with those faring best being the most capable of leading well. The accumulation of wealth and power is therefore not considered a bad thing, nor is the generational passing of this. There is also a belief, however, that with great wealth and power comes a responsibility act morally, provide leadership and alleviate the suffering of those with fewer resources – although their preference is usually for philanthropy or charitable assistance rather than state action.

As discussed in an earlier chapter, conservatives favour the economics of the free market, and this is closely linked to their attitude to property ownership. Since the later 19th century, the UK Conservative Party has drawn support and a not insignificant amount of funding from 'big business' – donations from rich individuals and corporations. Despite the so-called post-war consensus, the party opposed expansion of public ownership during the Labour governments of the 1950s and 1970s, although they tolerated significant state ownership during this period, and the Heath government of the early 1970s imposed price and wage controls in response to inflationary problems, even nationalising Rolls Royce, against all conservative instincts.

The Conservative predilection with the free market was put on steroids in the 1980s under the leadership of Margaret Thatcher. Her governments privatised many state-owned industries; many would argue based on ideology rather than pragmatism. Thatcher was heavily influenced by the free market philosophy of Friedrich Hayek and economic theories of Milton Friedman (2020). Friedman thought that even the relatively low level of intervention in the economy that J.M Keynes (1936) proposed was pointless in reducing unemployment instead produced inflation, undermined competitiveness, and ultimately corrupted the economic order. Their message was the economic and social excellence of the free market. Conservatives then, support private property rights and free trade, although this was not always the case. For example, at the end of the 19th century, many Conservatives supported 'imperial preference' in trade, rather than trading with countries outside the Empire.

More recently, there have also been significant strands of Conservative thought that question the current commitment to the free market. For example, conservative philosopher Roger Scruton (2001) challenges the idea that all could be reduced to the operations of the free market, whilst Michael Oakeshott (1991) and others sought to reconcile traditional conservative values with a championing of the free market. There are many conservatives who have been disappointed with the focus on the free market at the expense of more tradition conservative values (see Chapter 1).

Pre-eminent conservative philosopher, Roger Scruton, considers property rights central to conservatism although as he points out, is not defined by it. For Scruton, conservatism is a social theory and practice, with economics and ownership as integral but non-definitional aspects. Ownership is integral to our ability to conceptualise rights, responsibilities, duties and freedom (Scruton, 2001). Through the institution of property, objects cease to be mere inanimate things and become the focus of rights, obligations and social relationships. For Scruton then, the institution of private property is central to man's self-realisation as a social animal. Further, it enables the institution of *home* as the principal sphere of property, and the *home* as the basis of family, perhaps the key institution for conservatives.

For Scruton (2001) and conservatives more generally, home is the place where private property accumulates, and further with the family becomes some shared.

In the home, distribution is not the subject of any formal contract, rather sharing is "simply the essence of family life.=" (*ibid,* p. 94). Conservatives, therefore, see private property and the family as institutions that either stand or fall together. For them, family demands the establishment of property for its very existence. Scruton extends this out beyond ownership, stating that even households who cannot afford to purchase a home rely on property rights conferred on them via council tenancy agreements as the basis of their family life. The tenancy agreement can, of course, be terminated by a court in a way that a private lease cannot be, but nevertheless, the rights conferred by such an agreement enables the tenant(s) to accumulate and secure gadgetry and goods that transform the mere *house* into a *home*. Likewise, the ability to use communal facilities such as gardens, lawn mowers etc. is a conference of property rights that may transform the domicile into a personal experience. For Scruton (2001), the link between household and family is undeniable.

For many conservatives then, private property is not merely the basis of accumulation of monetary wealth or market transactions but the basis of their most revered institution, that of the family. It is the basis of all social relations and enables the accumulation of not only wealth as property value but also consumer goods and personal effects; tenants feel no less adamant in their defence of their home than the owner of any country house.

Another of conservatism's pre-eminent figures, Michael Oakeshott (1991), defends property rights on a different, yet no less compelling basis. Oakeshott is more up-front than Scruton in his view that property rights are a form of *power,* and further, institutional property is a way of organising and distributing this form of power across society. For Oakeshott, the different forms of property, personal and real, chattels, a man's own physical and mental capacity, are forms of power and spring from the same bases, investment, inheritance and luck. "Institution of property is unavoidable" (Oakeshott, 1991, p. 392).

Oakeshott (1991) also considers the institution of property rights as a bulwark against despotism. If we think back to our previous discussion on feudalism, the monarch owns all property and has the power to distribute or remove property from individuals at will. This gives one person virtually unlimited power over the rest of the population, having the ability to make a person either rich beyond their wildest dreams, or plunge them into dire poverty without hope of assistance. Property then is the arrangement most friendly to the concept of *freedom.* The best system being that which ensures that private property is distributed as widely as possible, and which discourages a dangerous concentration of power. As an example, Oakeshott (1991) shows that the right to own private property is self-limiting in that it proscribes slavery. "The right to own another man, could never be a right enjoyed equally by every member of society" (p. 393). As we have seen previously, there can be limits to property rights and in modern societies, these are generally instituted by democratically elected governments for some greater good. Take the recent controversy over compulsory purchase of land to build HS2. Many people have had to sell their businesses, houses, and other land to make way for this large infrastructure project. Many others will now have a major trainline running close to their property causing noise and visual disturbance. They have, of course, received compensation for these ills, and the fact that there are democratic and bureaucratic checks and balances mean that these have not been arbitrary decisions. Indeed, Oakeshott is clear that the institution of private property than best serves liberty is that which is subject to least arbitrary limits and exclusions. It is from this that the maximum individual power from ownership may be derived. As Oakeshott (1991) points out, have not the greatest threats of tyranny come from circumstances where the state, industry, corporations or trade unions have acquired exceptional proprietary rights?

King (2003) makes a similar argument to both Scruton and Oakeshott in that he sees housing and property rights and foundational to other rights and functions that we have and experience as human beings. These are what he terms 'freedom rights'. The argument for freedom rights in relation to homelessness are developed most effectively by Waldron (1993) when he argues that we cannot live any kind of meaningful life, if we do not have the freedom to carry out certain 'functions', such as sleeping, washing and toileting. For Waldron, property rules determine where one has a right to be, rights of use and exclusion (King, 2003). These tasks are all 'situated', meaning that they must be done somewhere. Waldron contends that a lack of a place to carry out these situated tasks is a breach of our rights. He discusses homelessness in its most obvious form, rooflessness, contending that one is not free to perform and action unless there is somewhere one is free to do so. He defines homelessness as the condition where one is excluded from all places governed by the rules of private property, homeless people are only free to be in public places, unless given permission by the owner to use their private property. Waldron presents the case for homeless people as the 'right to be free', to have personal freedom; and that means having access to a place to perform those tasks Waldron enumerates. This approach is extended by Martha Nussbaum (1999) when she argues that there are functions that are essential to human dignity and the right to these is inalienable. Nussbaum elaborates considerably on what these 'functionings' are. In short, she defines these as life and bodily health and integrity, senses, imagination and thought, emotions, practical reason, affiliation, other species, play and control over one's environment (Nussbaum, 1999; King, 2003). For Waldron and Nussbaum, housing is a means of ensuring good bodily health and integrity.

This approach is interesting as it lifts housing out of its usual context as a socio-economic and positive right, and places in into the realm of negative rights. That is to say that preventing a person from having a place to conduct their basic functionings is a breach of their negative right to be able to do so, thus living a worthwhile life. Second, having such a place to perform these functionings, means that we do not coerce others into suffering having to witness our functionings (Berlin, 1969), and we do not have to witness this of others.

Socialism, communism and private property

Although it is no less predictable, the attitude of these ideologies to private property is more interesting. As I covered in the chapter on ideology, socialism is a somewhat disputed term. Although it is sometimes seen as a synonym for full communism, it is also often seen as a stage on the way to full communism, or sometimes as an ideological position all its own. For example, in the UK, the Labour Party, Green Party and SNP and Plaid Cymru in Scotland and Wales respectively would call themselves socialist. This is not to say that the aspire to full communism (although some may argue that they do), but they do generally aspire to highly redistributive state with high taxes on wealth and property, and a generous welfare system that redistributes this wealth as cash benefits, subsidies or, of course, housing.

Where socialism is taken to be a stage on the road to full communism, it is often associate with Marx's 'dictatorship of the proletariat'. This is the stage immediately before full communism. As I said in Chapter 1, the writings of Marx and Engels are somewhat complex and truncated, but it is useful to understand their theories in summary here.

For Marx, the proletariat (working class) are exploited by the bourgeoisie (capitalist class) when the latter extract the excess value of the labour of the former to enrichen themselves and their class. In Chapter 3, we looked at economics and how value is derived. In short, the value of a good is a function of many variables, e.g. the cost of the materials used to make it, the cost of the labour used to make it, its abundance or scarcity in the market, quality etc. Marx disputes this. In his 'labour theory of value', Marx reduces value to a single factor, the labour used to produce it. For Marx, the more labour that is used to produce a good, the more valuable it is. Further, the whole value of the product belongs to the person that produced it. This leads to the inevitable position that when a capitalist uses the labour of a worker to produce a good, sells that good, pays the work less than the sale value and keeps the surplus, this is the root of capitalist exploitation. Further, Marx states that the bourgeoisie created a whole ideology, 'capitalism' to support this exploitation. In western society then, the whole social world is little more than a system to support this exploitation. Being able to buy consumer goods, religion – all hiding this exploitation. So essentially then, Marx divides the world into two classes, one exploiting the other. To change this situation, Marx called for a revolution, the role of the revolutionaries being to awaken a class consciousness in the proletariat such that they overthrow the bourgeoisie and take control of their labour. There are, of course, many people who disagree with Marx's ideas, including many of the people Marx says are exploited. Marx states that anyone who is not 'class conscious' (aware of their own exploitation) is suffering from 'false consciousness'. This is, of course, something of a Kafka-trap, and highly convenient for Marxists who can write off objections to their ideas as this 'false consciousness', without evidence that it even exists.

When enough people become class-conscious (aware of the exploitation), they take control of the 'means of production', be able to keep the full product of their own labour and at this point, everyone will be considered equal. Marx called this stage 'the dictatorship of the proletariat'. In this stage, the people will take control of the levers if state power and use them to enforce equality, or more accurately, equity (equality of outcome), and exploitation will end. Eventually, after living in this condition for some time, people will find it becomes second nature, the state withers away as it no longer necessary, and full communism will therefore emerge spontaneously and replace capitalism – communism is Marx's final stage of history. In full communism then, private property is abolished fully. No one can own anything; all resources will belong to the people. Socialism is often therefore, taken to be this 'dictatorship of the proletariat'.

For socialists and communists then, property rights are less significant than for conservatives or liberals. Rather than being the basis of a just society, ownership of private property is at the root of exploitation. It is therefore easy to see how for those more to left of politics, that taxation and other forms of redistributing income are much less controversial than for those on the right. Indeed, the appropriation of private property by the state is to be positively encouraged as the situation where the state owns all property on behalf of the people is positively to be encouraged. Remember that according to Marx's historicist approach, communism is inevitable and therefore all steps that progress towards it are to be welcomed.

As we have seen, many conservatives defend property rights first, on the grounds that they provide the ability to provide a home, and thus a place to raise a family, and second as a bulwark against tyranny by the state, corporations etc. In contrast, Marx's ideological soulmate, Friedrich Engels, writes that the family structure is a key site of oppression for women where they are socialised into passive social and political roles by virtue of their gender. For Engels, amongst others, the family should be abolished as child rearing revivified as a communal, societal activity thus freeing women and children from the tyranny of the family structure. On the second point, as I have already said, anything that provides a bulwark against state power is anathema to many on the left who see the state as key player in the development of the utopia.

Those on the left of politics also tend to be more egalitarian than conservatives and see the brining about of equality as a legitimate aim of the state. Private property and in particular, housing, is a key driver of wealth inequality. In extremis, communists value equity or equality of outcome above all other values. Private property, being a source of wealth storage is clearly inimical to this goal.

Whither housing policy?

As King (2003: 2011) rightly points out, it is all very well to discuss these highfalutin ideas in the abstract, it is all too easy to claim something as a right, but as Jeremy Bentham might have it, rights without a way to enforce them are 'nonsense on stilts'. Indeed, some might argue that housing cannot be a 'right' at all, given that it requires action from other parties to provide it – a positive right. However, as King (2003) argues, housing can also be seen as a negative right in that provides a place to perform the essential functions that make one truly free. This argument undercuts the socio-economic case for housing, relying instead on a moral imperative. In any case, the original concept of 'property rights' enshrined only the right to own property, but not the right to have that property provided by the state or any other person/body. As I have discussed previously, this may be fine in a frontier country (such as the US) but is not so effective in a country like the UK where there has been a long process of wresting power and property from the monarch to the people. Property rights, therefore, may be said to have expanded to have included the right to have that property provided and defended by the state. There is now a right to a standard of living, of which housing obviously plays a key part.

The statutory right to housing is enumerated in several national and international laws and treaties. Indeed, the right to be housed adequately has been recognised by the various governments of the UK for some years; even though these rights are sometimes qualified. Pre-existing much of the UK housing legislation is the Universal Declaration of Human Rights (adopted by the UN General Assembly in 1948). Article 25 of that declaration states,

> Everyone has a right to a standard of living adequate for the health and wellbeing of himself and of his family, including food, clothing, **housing** and medical care and necessary social services, and the right to security in the event of unemployment, sickness, disability, widowhood, old age or other lack of livelihood in circumstances beyond his control.
>
> *(Kenna, 2005, p. 1 [original emphasis])*

Whilst this may not be formally binding, there are several other instruments that do impose obligations on ratifying nations. Article 34(3) of the European Charter of Fundamental Human Rights states:

> In order to combat social exclusion and poverty, the Union recognises and respects **the right to social housing and housing assistance** so as to ensure a decent existence for all those who lack sufficient resources, in accordance with the procedures laid down by the Community law and national laws and practices.
>
> *(Kenna, 2005, p. 1)*

Kenna (2005) and Fitzpatrick and Watts (2010) also identify the UN International Covenant on Economic, Social and Cultural Rights, the Charter of Fundamental Human Rights and the European Social Charter as imposing housing related duties. Fitzpatrick and Watts (2010) state that perhaps most relevant is Article 31 of the Revised European Social Charter, which obliges ratifying states to "take measures designed to promote access to housing of an adequate standard, to prevent and reduce homelessness with a view to its gradual elimination and to make housing affordable to all" (Fitzpatrick & Watts, 2010, p. 109)

Under both international and European conventions, the right to housing is a protected right for which there is redress to the appropriate courts or other body for enforcement. One of the issues here may be the very use of the word 'housing'. As King (2016) discusses, the word 'housing' implies shelter, somewhere to live; but "home" is a slightly different concept. 'Housing' is what we use to make a 'home', a place to take shelter, to keep memories, to raise the next generation (King, 2016). Duties to provide 'housing' contained in the legislation and conventions discussed is a slight misnomer and what we should perhaps strive for is to provide homes. The provision of 'housing' alone does not in itself resolve 'homelessness', which should be reconceptualised to be about more than just shelter and to include the social interaction and concepts of memories and security that King (2016) discusses.

Governments in the UK have long been concerned with the provision of adequate housing particularly for groups that are deemed vulnerable. Legal duties have been conferred upon local authorities to provide both indefinite temporary and permanent housing to those owed a duty under several housing Acts. It is also clear that in some respects England has more robust legislative framework relative to many other European nations and that Scotland and Wales have yet more robust systems for dealing with homelessness. However, despite recommendations that tests of 'intentionality' and 'priority need' be dropped in favour of a universal duty to provide adequate housing, these tests remain in England and Scotland (such tests were removed in Wales by the *Housing (Wales) Act 2014*. In some respects, this is understandable, as local authorities must ration limited resources.

There are also European human rights-based arguments, the wording of both the United Nations (UN) and European conventions that state that housing is a human right and as such government has a duty to ensure that it is provided. Further, that this duty to provide adequate housing should also include a duty to provide the support that is necessary to maintain accommodation and further than that, to enable people to participate fully in civic and social life. These ideas suggest that housing be seen not as merely one of housing and private space (Edgar & Meert, 2005) but one of social inclusion; that vulnerability because of being poorly housed or lacking privacy also means vulnerability because of being socially excluded. This chimes with King's (2003) ideas about housing as a 'freedom right', and the basis for being able to perform those functions that can only be carried out in private.

The right to housing in the UK

The right to housing, defined in terms of social inclusion, in the UK can broadly be said to have begun in earnest with the inception of the welfare state following World War II and the subsequent Beveridge Report. Prior to this, housing for people who could not afford to buy was via the private rented sector or for the very poor, Houses of Industry or Correction, the workhouse, or 'lunatic asylums' (see Irving-Clarke [2016; 2019] for a full account). There were also pockets of charitably provided housing, form example almshouses provided by religious groups or the mercantile class (see Pannell & Thomas, 1999). After both world wars, there was determination to provide those who had been fighting on the front lines with a good standard of housing. This was largely delivered via a mass building programme by local authorities, council housing, or as it is now known, social housing. Although, the concept of 'social' housing has earlier roots.

The origins of social housing, as we think of it today, are commonly traced back to the early 20th century and Octavia Hill's efforts around the London slums. Although, some theorists see the provision of social housing as a method of social control rather than altruism. For example, crime control can be seen in relation to housing as a new problematisation of the housing crisis, adding that this has echoes of the foundations of housing policy in the 19th century and Victorian concerns around pauperised deviance (Cowan & McDermont, 2006). Watts and Fitzpatrick (2018) outline the use of welfare policy as a paternalistic tool to drive behaviour change especially since 2010, first by a Conservative/ Liberal Democrat coalition and then solely Conservative governments. They argue that these governments have pursued welfare policies that further residualise housing as a welfare provision as well as introducing a punitive welfare system designed to drive behaviour change in claimants. Using policy in this way has a long history see Irving-Clarke (2016; 2019), and Mead (2001) for example.

The 'right to housing' has its roots following a determination, after World War I, to provide 'homes fit for heroes' as they returned from the front (Yorke, 2017). The *Housing and Working Classes Act 1890* had empowered local authorities in London to build housing for poorer communities and charge reasonable, fair rents. Councils had to re-house a least half of the people displaced by these slum clearances. A decade later the *Housing of the Working Classes Act 1900* extended the 1890 Act to areas outside London and allowed councils to buy land outside their own districts. At the start of World War I nearly 90% of households lived in the private rented sector, with most of the remainder being owner occupiers (King, 2006). During the war, construction of new homes had been at a standstill as materials were directed to the war effort, leading to a shortage of urban housing and as a result rents increased dramatically. In response, the government introduced fixed rents and mortgage interest at 1914 levels via the *Rent and Mortgage Interests (War Restrictions Act) 1915*. The housing shortage was only made worse after the war ended as soldiers returned and started families (King, 2006). The resulting *Housing and Town Planning Act 1919* introduced a national housing programme that highlighted the shortage of good-quality homes for men returning from World War I.

The interwar years saw an acceptance for the first time that private enterprise would not be able to produce enough quality homes at affordable prices and that the state had a role to play in doing so (Burnett, 1993). The idea that the market could not provide homes for all led to a 'safety net' provided by the state to provide for those who were unable to access the market. The interwar years also saw an explosion in owner-occupation as wages rose in real terms, house prices remained static and the creation of building societies with high levels of liquidity allowed a stable, secure base for investment. Thus, the sector with the highest growth between the wars was owner-occupiers.

World War II had a dramatic effect on the provision of housing by the state. The government was faced with the two-pronged problem of a housing shortage (caused by bomb damage) and, again, returning soldiers and the related 'baby-boom' and growth in household formation (Irving-Clarke, 2016). The Beveridge Report (1942) introduced the concept of a comprehensive state safety net to protect its citizens from the five giant evils in society (squalor, ignorance, want, idleness and disease) from cradle to grave. The post-war Labour government, led by Clement Atlee, chose to meet the demand for new homes mainly via large scale local authority building programmes. House building numbers remained at over 150,000 units per annum right through the 1950s and were still above 80,000 per annum throughout the 1970s. It was in good financial shape with many older council properties were free of debt enabling reductions in rent or used to subsidise further building.

Despite the so-called post-war consensus the ideological debate about the state's role in mass housing provision and the encouragement of those who could do so to move out of social housing became twinned with social housing becoming less easily available. Local authorities were encouraged to allocate council housing on *need* rather than *desert* following the Cullingworth Report (1969) meaning that the essence of social housing was changing. Mullins and Murie (2006) point out that council housing had always been provided to those who could afford the rent, e.g. the more affluent working class and that those who could not afford to pay were residualised in the sub-standard private rented sector (or worse). In 1979, 20% of families in the top decile of income distribution lived in social housing, compared to almost zero by 2004/05. Until this point, council housing had been an aspirational tenure for the relatively wealthy working class (CIH, 2018).

Prior to the 1972 Act, local authorities did have the power to charge a subsidised rent on a discretionary basis but as Cowan and Marsh (2018) point out, the lack of security of tenure (only introduced in Right to Buy legislation during the 1980s) left households vulnerable to eviction at the whim of local authorities; and withdrawal of subsidised rent could be used as a lever to produce a desired result, e.g. moving to a smaller property in cases of under-occupation (CHAC, 1953, p. 14). Although Cowan and Marsh (2018) note that this was not common practice.

Beginning with Octavia Hill and continued via local authority provided housing after the two world wars, the provision of and right to, good-quality, affordable housing was via rented housing, in the case of both Hill and council housing, the provision of this housing was in exchange for a modification of conduct. Hill demanded that tenants keep properties in good condition and not indulge in anti-social behaviour, likewise local authority housing departments demanded the same. Tenancy handbooks from the time demand houses were kept clean and gardens tidy. As noted in the previous paragraph, council tenants had no security of tenure and so could be easily removed from council properties if they did not conform to the standards expected. This changed in the early 1980s.

The right to 'own' housing

The election of a Conservative government under Prime Minister Margaret Thatcher in 1979 altered the course of housing policy permanently. As an adherent of Friedrich Hayek and Milton Friedman, Thatcher emphasised a commitment to market provision of housing, promoting a revival of the private rented sector and a belief in extending home ownership with social housing being aimed only at those who could not compete in the market. The Conservative's focused on setting people free from government constraints on their lives by rolling back the state. This included removing the power local authorities had over tenants by giving them security of tenure and giving them the right to buy their homes.

The monetarist policies of Milton Friedman prescribed limited state intervention in a 'self-correcting' free market-drove housing policy during the 1980s and well beyond (Irving-Clarke, 2016; 2019). Gough (1980) argues an ideological attack on welfare was firmly cemented by the Thatcher governments; although he suggests this was largely initiated by the Labour governments of Wilson and Callaghan in 1975. He further argues that the qualitative shifts in social policy were aimed at reasserting individualism, promoting self-reliance and family responsibility, and to ultimately dismantle the ideology of the collective social provision of the post-war era (Henderson, 2019).

Driving policy was the Government's preference for home ownership over other tenures depicted in the *Housing Act 1980* (see Gurney, 1999) which introduced the controversial, flagship 'Right to Buy' policy. Local authority tenants could now purchase their property at a discount dependent on how long they had been a tenant. The policy proved extremely popular with voters and has never been fully repealed by successive governments of either party in England; although the *Housing Act 2004* introduced stricter regulations including a tighter time frame in which tenants could sell after purchase and with less discount to buy. The Scottish government and the Welsh Assembly have both ended it in a bid to protect social housing stock from further reduction and ensuring the state can provide safe, secure accommodation to those who are unable to buy or rent their own home.

The effects of the 'Right to Buy' policy were manifold and widespread; it removed much of the most desirable and sought after stock from the social housing sector, without a replacement building programme (Forrest & Murie, 1988). This point is borne out by Manns (2017) who stated that since the introduction of the policy, it is estimated that over two million properties have been moved into private ownership. Lund (2011) argues that Right to Buy and other policies resulted in a continued concentration of low-income households in the social housing sector with those accessing social housing increasingly in the greatest housing and social need. Social housing was becoming more synonymous with welfare and a clear polarisation of the housing market was established. Hills argues that housing policy hardly existed anymore but survived within the welfare state, "mainly as an adjunct of social security (through Housing Benefit) or as part of wider city regeneration policies" (Hills, 1998 p. 13 quoted in Tunstall, 2015).

However, King (2006) argues that 'right to buy' is one of the most successful housing policies there has been. This is largely because it delivered against its stated aims. There is a considerable academic and grey literature on policy implementation questioning why policies succeed or fail, and what can be done to make better, more effective policies. In terms of meeting its policy aims, right to buy largely succeeded as a policy, because it meets these aims. Indeed, when right to buy is criticised, this is largely done on the basis of criteria the policy was not meant to achieve. King (2006) compares right to buy with the 'Tenants' Choice' policy of the same period. 'Tenants' Choice' would have empowered social tenants to vote to transfer their tenancies to a different landlord. The point of this policy was to encourage competition between landlords, the best would clearly expand their stock as tenants voted to join them. However, the policy was a failure. King (2006) contends this is because 'Tenants' Choice' offered no change in the control of the property to the tenant, only a change of landlord. Right to buy on the other hand offered tenants to take full control over their home by enabling them to purchase it. Thus, this policy succeeded where 'Tenants' Choice' failed.

The Conservative governments of the 1980s and 1990s continued the doctrinaire repositioning of state-controlled housing out of the government arena via the *Housing and Planning Act 1988* which gave councils the option of transferring all or part of their housing to another landlord, such as a Registered Social Landlord (RSL). The 1988 Act allowed council tenants (as a whole) to choose to transfer their existing homes to another landlord, either a separate housing association (LSVT) or ALMO operated by the local authority but managed autonomously.

The election of a Labour government (under Tony Blair) in 1997 saw some change in the fortunes of social housing. Although they continued the policy of seeing home ownership as the pinnacle of housing, but also initiated a major investment in the social housing stock under the Decent Homes Programme, invested in housing-related support via the Supporting People programme and made major inroads into tackling homelessness and rough sleeping. However, the Thatcherite 'right to buy' policy continued as did LSVTs and creation of ALMOs to transfer housing stock out of direct local authority control. These measures further residualised social housing as a tenure for those on the lowest incomes (CIH, 2018). Pawson and Mullins (2010) argue that New Labour perceived council housing as a redundant project with low demand, poor quality stock and saw the issue through the lens of poverty and limited life chances and opportunities, not merely as a matter of giving tenants more choice of landlord. Labour recognised there was a stigma attached to the widening difference between homeowners and those living in estates with low demand and poverty, with the term 'sink' estate becoming used by many commentators. New Labour did, however, recognise the need to improve the housing product on offer. The Housing Green Paper, "Quality and Choice: A Decent Home for All" (DETR, 2000) aimed to bring all social housing into decent condition by 2010. New Labour reflected the Conservative approach in that it saw the state in an enabling role in relation to housing as opposed to necessarily providing and managing it.

This shows that the nature and role of social housing is not static and has changed significantly over time and space. Malpass and Victory (2010) suggest the changes have been from "public" to "social" housing (p. 7), from municipal to residual, local authority ownership with local democratic accountability to managerialist with heavy regulation from the centre. Alongside this, Fitzpatrick and Watts (2018) argue that social housing has moved from being a secure tenure for a range of households to an "ambulance service" for those in the most need or having been 'welfare-ised'.

Alongside these changes, I would argue that there has also been a changing attitude to the 'right to housing' along ideological lines. The two world wars in the early 20th century had profound effects on housing policy, with both being followed by significant regeneration and house building programmes to provide homes fit for soldiers returning from war. At these times, the right to decent-quality housing was clearly at the forefront of policy. It may not have been framed as such specifically but was rooted in gratitude for the sacrifices made in the service of country and against the forces of tyranny. Particularly following World War II, there was a determination that the state should provide enough high quality, affordable housing for all, although in the event it did not quite extend to those on the lowest incomes. This 'right to housing' extended across several other policy areas with the inception of universal state provided healthcare and education also.

The election of a Conservative government under Margaret Thatcher in 1979, which has some radically different ideas about the role of the state, saw the inception of the 'right to buy' your council home as a means of freeing people from state interference in their lives. Earlier in the chapter I reflected on the ideological differences between left and right and how this affected housing policy. Whereas the post-war consensus governments, of left and right, saw housing through the lens of having a decent, affordable place to live, Thatcher's Conservatives saw the issue through a new lens, that of the right to be free of state interference, with housing (and other things) being better provided through the market. Although, there was still acceptance of the role for state housing, but as a 'safety net' for the poorest, rather than a service for all who needed/wanted it. Overall, ideas about a 'right to housing' have become much reduced since 1979 with both Labour and Conservative governments since, pursuing 'right to buy' as a key housing policy, and the use of social housing as a welfare safety net for the poorest households (King, 2016).

The election of a (new) Labour government in 1997 saw a partial revival of some aspects of social housing policy, with the Supporting People programme to provide support for the most vulnerable, and Decent Homes Programme which provided huge investment in social homes and set a quality standard for social housing which had been lacking for many years. The right to a decent quality home was restored via this, as minimal as some thought the standard to be. The 'right to home' was therefore expanded to include a home of a defined, specific standard.

The standard of social housing was once again thrown into the spotlight in 2017 by the tragic fire at Grenfell Tower in North London. The fire highlighted not only that the tower block had been covered in dangerously flammable cladding but also that appropriate checks had not been carried out, fire doors and other essential safety equipment was missing, and the 'stay put' policy was woefully inadequate. Later in around 2021, ITV news uncovered widespread disrepair on the Eastfields Estate, also in London and owned and managed by a social landlord, Clarion Housing Group. At Grenfell Tower, the residents had raised a string of concerns about fire, and wider, safety in the block; the tenants on the Eastfields Estate had also repeatedly reported disrepair in their homes. These issues raised serious questions about the way that social tenants are treated both by their landlords and more broadly. The stigmatisation of social housing and tenants has been an issue in the sector for some years, with campaign organisations such as See the Person and the Chartered Institute of Housing taking up the cause in recent years (CIH, 2020). In response, the Conservative government published both a Social Housing Green Paper (2017) and Social Housing White Paper (2021) that promised to both improve conditions in social housing and the way that landlords and the wider public view social housing tenants.

A new tripartite regulatory arrangement for social housing is now in place. The RSH has published new, stronger principles it will apply in regulating the sector, and the draft outcomes it will be assessing for. Once the government passes the 'Social Housing Regulation Act', giving shape to the role of the regulator, there will be further consultations on the regulatory standards and how the inspection regime will operate. RSH is already consulting on the proposed 'Tenant Satisfaction Measures' that will be used to assess landlords. At the time of writing, we are awaiting the publication of these measures and their passage through Parliament.

On building safety, the *Building Safety Act 2022* had recently completed its passage through Parliament at the time of writing. In summary, the bill creates a clear, proportionate framework for the design, construction, and management of safer, high-quality homes, and strengthens the regulatory regime around construction products to ensure they are safe. Building owners will be required to manage safety risks with clear lines of accountability for safety during design, construction, completion, and occupation. Those who don't meet their obligations under the proposed Act may face criminal prosecution. The Building Safety Bill introduces a new regulatory framework for high-rise buildings, including:

- The HSE becomes a statutory consultee for planning applications for high-rise buildings
- The BSR will become the new control authority for high-rise buildings
- A requirement for a 'Golden Thread' of building information – identified, updated and stored throughout a building's lifecycle
- Mandatory reporting of prescribed fire and structural safety occurrences.

The Bill also introduces the 'Accountable Person'. This will be the organisation or person who owns or has responsibility for a building, or who is responsible for maintaining the common parts of the building. The 'accountable person' will have a duty to take all reasonable steps to:

- Prevent a building safety risk happening – this is defined as 'spread of fire and/or structural failure'
- Reduce the seriousness of an incident if one happens.

The Housing Ombudsman Service is already largely operating under its new remit, is publishing a wide variety of reports, data, and training materials. More broadly, the Future Homes Standard continues to shape environmental standards for new homes, with the building regulations being revised to support its goals. An updated Decent Homes Standard, proposals for electrical safety checks in social housing, vehicle charging points in new homes, and greater protections for private renters are also forthcoming.

Embedded throughout the Social Housing White Paper, professionalisation and stigma overarch many of the other themes. The government has convened a working group to review the training and staff development of social housing staff. The Chartered Institute of Housing have launched a professional standards framework for people working in housing, and it seems likely that this may form a key part of any new approach to professional development coming out of the working group.

These changes following Grenfell Tower and Eastfields scandals expand the right to housing yet further. What began as merely the right to own and control private property, has now expanded to the right to have that property provided by the state or some other entity, and then to a home that meets a specific set of standards; these standards are now being gradually expanded to include all manner of features and amenities. This is not necessarily a bad thing, who would disagree that homes should meet basic safety standards, and be weatherproof? At the last general election in 2019, the Labour Party manifesto (Labour Party, 2019) promised free broadband in all homes, provided by the state. As the Conservatives won the election, this was never delivered, but this does raise the question of what the limits of state provision should be. Should the state provide broadband? What might happen if a less than liberal government came to power and could control people's internet access? We only need look at China to see where this might end up.

Concluding comments

This chapter opened with a broad discussion about property rights, what they mean and how they came to exist. There is a clear antagonism between the left and right of politics over the right to own property, particularly with reference to housing. As I have discussed classical liberals and conservatives see the right to own private property as a bulwark against state power. Housing being a source of wealth and security, and an institution against what they see as the predations of an overmighty government. This right is not total; however, even conservatives see the benefits of the state being able to commandeer property for its ends, such as infrastructure projects. However, for conservatives, the right to to do this is again constrained by democratic institutions of parliament and the courts, holding the executive is check with the constitution and its own laws.

The left sees property ownership as much less about this, and more about entrenchment of wealth by the upper classes with housing and other property used as a store of wealth rather than as a bulwark against state overreach. In terms of housing itself, the left see the provision of state or quasi-state-owned housing as a way of enabling better housing conditions amongst those who either cannot or do not want to access the market. Bevan's original vision was that people from all walks of life and levels of wealth would live together in state owned housing.

In the latter parts of the chapter, I discuss in more detail how the ideological differences discussed earlier have bled through into actual policy, before discussing housing conditions and the right to live in good quality housing beyond merely having shelter. New Labour revivified this from the earlier Labour policies by putting in place the Decent Homes Standard, to ensure social housing is weather tight and of liveable condition. This is itself in the process of being reviewed by the current government following the fire at Grenfell and revelations from Clarion's Eastfields Estate. The right to housing then is expanding to encompass standards of decency and safety in the private and social rented sectors, thus expanding the view of property rights beyond merely the 'right to own'.

So, can I burn my house down if I feel like it?

To go back to the opening of the chapter, it seems obvious that you are free to dispose of your possessions in the way you like, and this is a key characteristic of property rights. The especially applies to smaller items such as clothes, books, and other effects. However, as I have said elsewhere, housing is different. It is usually a large asset, it is a valuable asset, and it is designed to last a long time. Indeed, some houses are hundreds of years old.

That said, despite the rights conferred on renters by tenancy agreements, they would clearly not be legally able to burn their rented house down. This is not a right so conferred by the tenancy agreement. The ultimate owner may hold this right though. Even as the outright owner of a property, it still may not be lawful to do this. For example, the owner of a house which is detached from others and in the middle of a field may well be able to claim the right to burn that house down because there are few, if any, externalities to that course of action. It does not affect anyone else. However, if the houses were in woods, or part of a terrace of houses, or even attached to only one other property, there would be no right to burn the house down as damage to the houses attached or surrounding features would be inevitable; and we would not have the right to damage these things. This is where we may reach for Nozick's (1974) concept of 'side restraints', where we can see that actions that negatively affect others cannot be allowed, event by libertarians.

So, rather unsatisfyingly, the answer is 'it depends'.

References

Aquinas, T (2014) *Summa Theologica*, Catholic Way Publishing [Online].

Bentham, J (1987) *Utilitarianism and Other Essays*, Penguin, London.

Berlin, I (1969) *Two Concepts of Liberty: Four Essays on Liberty*, Oxford University Press, Oxford.

Beveridge Report (1942) *Social Insurance and Allied Services*, HMSO, London.

Burnett, J (1993) *A Social History of Housing 1815–1985*, Routledge, Abingdon.

CHAC (1953) *Transfers, Exchanges and Rents*, HMSO, London.

CIH (2018) *Rethinking Social Housing*, CIH, Coventry.

CIH (2020) *It's Not Okay*, CIH, Coventry.

Cowan, D & Marsh, A (2018) A perennial problem? On under occupation in English council housing, *Housing Studies* 34(2) pp.319–337.

Cowan, D & McDemont, M (2006) *Regulating Social Housing: Governing Decline*, Routledge-Cavendish, Abingdon.

Cullingworth Report (1969) *Cullingworth Committee Report on Council Housing: Purposes, Procedures and Priorities*, HMSO, London.

De Soto, H (2000) *The Mystery of Capital: Why Capitalism Succeeds in the West and Fails Everywhere Else*, Black Swan, London.

DETR (2000) *Quality and Choice: A decent home for all*, HMSO, London.

Edgar, W & Meert, H (2005) *Fourth Review of Statistics on Homelessness in Europe*, FEANTSA, Brussels.

Fitzpatrick, S & Watts, B (2010) The right to housing for homeless people in E O'Sullivan, V Busch-Geertseema, D Quilgars & N Pleace (eds), *Homeless Research in Europe* pp. 105–122, FEANTSA, Brussels.

Fitzpatrick, S and Watts, B (2018) *Welfare Conditionality*, Routledge, Abingdon.

Forrest, R & Murie, A (1988) *Selling the Welfare State: The Privatisation of Public Housing*, Routledge, Abingdon.

Freeden, M (1991) *Rights (Concepts in Social Thought)*, OU Press, Maidenhead.

Friedman, M (2020) *Capitalism and Freedom*, University of Chicago Press, Chicago, IL.

Gough, I (1980) Thatcherism and the welfare state: Britain is experiencing the most far-reaching experiment in new right politics in the western world, *Marxism Today* pp.7–12.

Gurney, C.M. (1999) *Pride and* prejudice: Discourses of normalisation in public and private accounts of home ownership, *Housing Studies* 14(2) pp.163–183.

Henderson, K (2019) *The Role of Housing in a Coordinated Response to Domestic Abuse*, Unpublished PhD thesis, University of Durham, Durham.

Hills, J (1998) *ESRC Research Centre for Analysis of Social Exclusion: Annual Report 1998/9*, ESRC, Swindon.

Hobbes, T (2014) *Leviathan*, Wordsworth Editions, Stansted.

Irving-Clarke, Y (2016) *Supporting People: How Did We Get Here and What Does it Mean for the Future?* Unpublished PhD Thesis, DeMontfort University, Leicester.

Irving-Clarke, Y (2019) *Supported Housing: Past, Present and Future*, Routledge, Abingdon.

Kenna, P (2005) *Housing Rights and Human Rights*, FEANTSA, Brussels.

Keynes, J.M. (1936) *The General Theory of Employment, Interest and Money*, Palgrave MacMillan, London.

King, P (2003) *A Social Philosophy of Housing*, Ashgate, Aldershot.

King, P (2011) *The New Politics: Liberal Conservatism or the Same Old Tories?* Policy Press, Bristol.

King, P (2016) *The Principles of Housing*, Routledge, Abingdon.

Labour Party (2019) *It's Time for Real Change: The Labour Party Manifesto 2019*, Labour Party, London.

Locke, J (2016) *Two Treatise of Government*, Createspace, Scott's Valley CA.

Lund, B (2011) *Understanding Housing Policy*, Policy Press, Bristol.

Machan, T.R. (1989) *Individuals and Their Rights*, Open Court, Peru IL.

Malpass, P and Victory, C (2010) The modernisation of social housing in England, *International Journal of Housing Policy* 10(1) pp. 3–18.

Manns, J (2017) The housing crisis will only get worse until England scraps right to buy, *The Guardian*, Oct 11 2017.

Mead, L (2001) *Beyond Entitlement: The Social Obligations of Citizenship*, University of Michigan Press, Ann Arbor.

Mullins, D & Murie, A (2006) *Housing Policy in the UK*, Palgrave MacMillan, London.

Nozick, R (1974) *Anarchy, State and Utopia*, Wiley-Blackwell, Hoboken NJ.

Nussbaum, M (1999) Women and equality: The capabilities approach, *International Labour Review* 138(3) pp.227–245.

Oakeshott, M (1991) *Rationalism in Politics and Other Essays*, Liberty Press, Indianapolis IN.

Pannell, J & Thomas, C (1989) *Almshouses into the Next Millennium: Paternalism Partnership, Progress*, Joseph Rowntree Foundation, York.

Pawson, H & Mullins, D (2010) *After Council Housing: Britain's New Social Landlords*, Palgrave McMillan, London.

Scruton, R (2001) *The Meaning of Conservatism*, Palgrave, Basingstoke.

Tunstall, B (2015) *The Coalition's Record on Housing Policy, Spending and Outcomes 2010–2015*, LSE, London.

Waldron, J (2003) Security and Liberty: The Image of Balance, *The Journal of Political Philosophy* 11(2) pp. 191–210.

Watts, B & Fitzpatrick, F (2018) *Welfare Conditionality*, Routledge, Abingdon.

Yorke, T (2017) *Homes Fit for Heroes: The Aftermath of the First World War 1918–1939*, Countryside Books, Newbury.

6

WHY CAN'T GOVERNMENTS JUST DO WHAT THEY LIKE ONCE THEY GET ELECTED?

Case study

Governments are elected on the basis of the electoral system in the country that they will govern. Whilst these can be controversial, for example the 'First past the post' system in the UK, most are accepted as generally legitimate by the public and wider opinion. To win an election, the government must win some kind of majority of votes from the public. This might be a 'popular vote', whoever gets the most votes wins; or might be mediated by some kind of constituency system, e.g. in the UK or in the US in congressional elections. Staying in the US, the presidential election is conducted using the 'electoral college', a system designed to protect less populous states from the 'tyranny' of large, populous coastal states that tend to contain large cities.

Governments are also elected on the basis of a manifesto that they put forward as part of their election campaign, and states what they will do once elected. These are usually relatively ideological documents that accord strongly with the parties' ideological positions. Although this can vary considerably.

Once elected with a significant majority of seats in the House of Commons, the UK government can get pretty much anything it wants to do through this part of the process. The *Parliament Act 1911* places limits on what the House of Lords can do to hold up or stop government legislation, particularly if it is implementing a manifesto promise. The final part of the process, Royal Assent, is a convention and formality. The monarch could not refuse to give their assent without provoking a constitutional crisis.

DOI: 10.1201/9781003298212-7

Yes, governments cannot do anything they like. In previous chapters, I have discussed at length, some of the formal and informal limits they must work within. But, what if there are other factors limiting what governments can do?

- What might some of these limits be?
- Can you think of any recent, or not so recent examples, of factors or events that constrained government action?
- Are there some events that do the opposite and open windows for policy action that were previously closed?

Introduction

In previous chapters, I have discussed some of the theoretical and ideological concepts behind political parties and their (housing) policies. Many of these theories start from a 'blank slate' position (Locke, Rawls, Nozick etc.), assuming that nothing precedes the developments they detail. In the case of the US, this may be largely true as the founding fathers found themselves in position to start from scratch on the country's constitution. They did have to deal with the existing, native population and the legacy of British colonial rule, indeed the source of much of American jurisprudence originates in the British common law, and their system is based on many of the ideas of a British philosopher, John Locke.

The context that we are discussing here, the UK and latterly, English law is not so simple. As I hope I have detailed in the earlier chapters, the UK has a considerable history of both tradition and institutions that means change can be hard to achieve. An ultimately (small 'c') conservative establishment constrains the ambitions of radicals, the law, the church, the courts and conventions all act as constraints on the level of change any government can affect. This can be seen most recently in the considerable resistance several governments have faced in implementing the referendum decision to leave the EU – Brexit. Despite the controversy, I hope it is not too controversial to say that there has been considerable resistance to Brexit from the Civil Service, politicians who campaigned to remain and others, who have mobilised their institutional, political and cultural power to either prevent or water down the UK's exit from the EU.

It would also be remiss here, not to discuss the role of expertise in the policy process. A plethora of university academics, think tanks, civil society organisations and others seek to influence the policy process. Some are more obviously ideologically motivated, e.g. Trade Unions, whilst others have at least a veneer of independence attached to them. Over recent decades, academics have been increasingly encouraged/required to demonstrate that their research has an impact on society at large. Some scholars trace this 'impact agenda' back to the 1960s when the idea of the 'disinterested', 'value-neutral' expert lost its plausibility, meaning a new, more explicit, and legalistic social contract had to be drawn up under which scientists were required to demonstrate that their research was relevant and beneficial to society (Sarewitz & Pielke, 2007; Allen & Imrie, 2010).

Others link it to the rise of neoliberalism, and the privatisation and corporatisation of the UK academy (Smith, 2010). Still others highlight the rise of the 'evidence-based policy making' regime under the New Labour government which aimed to use science as a means of transcending ideological disputes. (Cabinet Office, 1999; 2000; Blunkett, 2000; Nutley et al., 2007). Regardless of the origin, there can be little doubt of the overall trajectory. The rise and evolution of the impact agenda has provoked a longstanding debate about whether the academy does enough to disseminate its research to policymakers and the public.

Academic researchers can suffer a conflict between offering practical advice and authoring scholarly articles for journals (from which impact is usually measured). The impact of social science research can be characterised in the following ways. The 'instrumental', influencing the development of policy, practice or service provision, shaping legislation and/or altering behaviour; and the conceptual, contributing to understanding of these, reframing debates and capacity building, e.g. through technical or professional skills (Nutley et al., 2007; Armstrong & Alsop, 2010). Meagher and Lyall (2013) add two more subtle ways that research can have an impact; cultural change, e.g. increased willingness of people and organisations to exchange knowledge or changed mindsets; and enduring connectivity, for example the establishment of improved relationships or partnerships between academics and practitioners.

It would be gratifying for politicians of all stripes, academics and others to think that there is a straight line between their ideologies, policy ideas, and reports and manifestos and the policy formulation and implementation process. However, this is simply not the case. There are omnibus other factors that influence the policy making and implementation processes, bringing change. Political concerns, path dependencies, exogenous shocks and other events can all have a significant bearing on policy decisions at any given time, making some changes possible and limiting the desirability and viability of others.

The thesis of this chapter is that for all the storm and thunder around ideology, there are other factors that arguably have greater impact than this, the academy and its research. Historical events that are largely exogenous to either research (experts) or policy are at least, if not often more, impactful than experts. Further, when we talk about research and impact, we often only talk in terms of the positive impact, e.g. how services for people experiencing homelessness, or housing space standards have improved, but there have been times when the impact of experts has been positively deleterious. This chapter explores some of these alternative perspectives.

How do governments get what they want?

Governments get what they want through implementing policies. These are implemented (and changed) for several reasons, maybe to keep things as they are, but more often to effect change to individuals/groups, or to change environments (Drake, 2001); of course, doing nothing can equally be considered a policy. In the UK, ideology plays an important role in policy change; blessed, or cursed, as we are, with not only two main political parties that are (mostly) ideologically opposed but also an adversarial political system. This is not to say that these left/right positions are in any way fixed and there are times when the ideological space between the Conservative and Labour parties narrows and widens. In his work that helped undergird New Labour's 'third way' project, Giddens (1998) identifies a retreating role of left and right in UK politics and in society generally (at the time). This has since been somewhat reasserted.

Giddens proposed (correctly) that left, and right are continuously in flux; what is considered left wing at one point in time might be considered right wing at another. ('Left' here is used as shorthand for more egalitarian, pro-state policies, whilst 'right' is used to connote more conservative, small state policies.) Bobbio (1994) agrees that as a necessarily adversarial process, politics needs left and right; two sides to oppose one another. He hypothesises that there are periods where left and right are clearly distinct, countered by periods where their policy programmes are very similar. Giddens (1998) gives the post-war period (1945–1979) as an example of this. Following the fall of fascism, the right in Europe accepted a good deal of the left's egalitarian aims and policies resulting in an enduring post-war political consensus. Likewise, the hegemony of neo-liberalism during the 1980s and 1990s forced the left to accept many of these policies. Finally, there will be times when the divisions between left and right reassert themselves (Bobbio, 1994), characterised recently with the Labour Party's left-wing egalitarianism (under Jeremy Corbyn) against a more right-wing Conservative party under David Cameron, Teresa May, and Boris Johnson. Labour is now seemingly tacking back to the centre under leader, Sir Keir Starmer; whilst the Conservative Party has been forced into some high spending policies, first by the COVID-19 pandemic and now the 'cost of living crisis' caused by the aftermath of the pandemic and Russia's invasion of Ukraine.

Ideology offers one possible explanation for policy change, here I will briefly look at some others before moving onto some examples that illustrate my point. Theories around 'path dependency' offer some alternative reasons, largely outside ideological concerns, but not totally disconnected. Previous events and policies can impose significant constraints or costs on present choices, making policy change less likely (Wilson, 1989 cited in Pollitt, 2008). There are situations in which events of the past are crucial to the feasibility of policy change in the present.

Pollitt (2008) gives the extreme example of 'irreversibility', circumstances where the costs of change or reversal are so high, it makes change unviable. Malpass (2011) outlines a similar strong version of path dependency, where it is implicit that the more steps one takes along a given path; the harder it is to return or change direction. Pierson (2000) uses the concept of 'increasing returns' to illustrate this point. This could be the increasing return on an investment of time, money or effort or represent the increasing costs of changing course. This is a useful concept in that it makes the point that once a particular path has been embarked upon it can be difficult to go back. However, the increasing returns argument struggles to explain how periods of change are precipitated. Rather than getting stronger over time, most policies fail; it is therefore the diminishing returns of a policy that make change more likely and not less (Marsh & Rhodes, 1992).

Strong path dependence, therefore, may be useful for looking at entrenched institutions such as government, churches, charities etc., but policy is much less entrenched; it should be treated with caution when dealing with the analysis of policy (Bengtsson & Ruonavaara, 2010; Malpass, 2011). Streeck and Thelen (2005) argue that change can often result from an accumulation of gradual, incremental changes and suggest several ways in which change can occur without the need for shock, whether internal or exogenous. First, the slowly rising salience of subordinate institutions to usurp the dominant (displacement), new elements may attach to existing institutions gradually changing their status and structure (layering), neglect of institutional maintenance in spite of external change results in slippage in institutional practice on the ground (drift), a redeployment of old institutions to new purposes whilst still attached to old structures (conversion) or a gradual withering away of institutions over time (exhaustion).

Bengtsson and Ruonavaara (2010) offer a weaker concept of path dependency where "one event, which is more or less contingent, considerably changes the probability of subsequent events or outcomes" (p. 195). This widens the scope of path dependency from one of simply cause and effect (the strong version) to a situation where events can be contingent, but still limit future options; and makes this a matter of degree rather than an either/or equation. They also go on to make three points about path dependency that underpins their own analysis (of Nordic housing). First, that there is a possibility that:

> single events, which are not the product of larger social forces or trends, might be influential to societal outcomes. This could even involve some element of chance. Secondly, previous events might be distant in time to the outcomes that are explained by them. Thirdly, as the process is 'essentially historical', it can only be analysed historically, that is by paying special attention to the temporally ordered sequence of events that leads to the outcome.
>
> *(Bengtsson and Ruonavaara, 2010, p. 195)*

Bengtsson and Ruonavarra (2010) make a key point here. Namely, that whilst past decisions and policies can restrict our policy options in the present, events also matter. Further, these events might be relatively small, contingent ones, or they may be larger cataclysms that have profound effects and consequences. Pollitt (2008) directly contrasts two different types of policy change, deliberative/learning based (the territory of the expert), and the forced/unexpected (exogenous events), adding:

> The relative frequency of these two types of path switch (deliberative/learning-based versus forced/unexpected) is a fascinating and largely unexplored question.
>
> *(Pollitt, 2008, p. 45)*

Here, I also wish to draw upon John Kingdon's seminal work 'Agendas, Alternatives and Public Policies' (Kingdon, 1995), as it will help us to understand the circumstances around policy change in terms of politics. Kingdon theorises that there are three major 'flows' running through any government and public life more generally, the flow of problems, the flow of policies and the flow of political activity. When governments can connect all three flows, major policy change becomes possible. Kingdon (1995) terms these rare confluences 'windows of opportunity'. Further, he identifies two types of windows. First the 'problem window', where a particular problem is seen to be fashionable or the public is clamouring for action, and second 'policy' windows, where opportunities open because of the appearance of a compelling problem or activity in the political stream. Kingdon also refers to the existence of 'policy entrepreneurs' who may agitate to create such windows and put pre-prepared policies forward to secure the changes they desire – or 'time tactics' (See Pollitt, 2008). Pollitt (2008) and Kingdon (1995) also speculate (probably accurately) that there is a "primeval soup" (Pollitt, 2008, p. 128) where different policies float around, bumping into one another and meeting new ideas, waiting for a suitable 'window' to open. This is perhaps best known as the 'Overton window', a conceptual device for conveying the political acceptability of a given set of policies at any one time.

What I wish to propose here is that for all the ink that has been spilled in the study of housing policy, notably Mullins and Murie (2006), Malpass and Murie (1999), Balchin (1995), Malpass (1990), King (2016), numerous journal articles and publications by the Chartered Institute of Housing (including the annual UK Housing Review) and National Housing Federation, large and small exogenous events have had significant impact and propelled change in housing policy, both outside the role of experts and political ideology. Although, they may have opened windows enabling the implementation of ideological policies or closed them off.

Housing policy is under theorised generally (King, 2010). This makes it an ideal case study to use here as it will not be complicated by other theories. Housing policy affects us all, is an area which is well-documented – there is a well-developed historical literature on the subject which sets out the history of housing policy in exhaustive detail. Lastly, it is a policy area where an, at least, rudimentary history is already in the public consciousness. Pollitt (2008) contends that one of the best methods for analysing such questions is to construct a historical narrative, that where I turn to in the following section.

(Social) Housing Policy

As I have said earlier in this book, (Social) housing policy has a relatively short history in this country with much of it constructed as part of the post-war consensus welfare state (c.1945–1979). There were also pockets of charitably provided housing. Otherwise, housing was largely a mix of owner occupiers and private renters. Also, many people lived in accommodation that was tied to their job, e.g. housing provided by landowners for farm staff or later in worker villages such as those pioneered at Bournville by the Cadbury family.

Shock one – inflation, and a decline of traditional methods of support

Following the introduction and incremental progress of the early poor law, mainly as relief for those unable to work, significant reforms were made during the Elizabethan period. The policy window was opened, and the changes largely driven by a serious deterioration in the economic circumstances of the country, combined with a decline in traditional sources of support for the poor (Boyer, 1990). The country experienced rapid inflation caused by rapid population growth, debasement of the coinage in 1526 and again in 1544–1546, and a significant influx of American silver (Boyer, 1990). As a result, the price of grain and other commodities increased significantly between 1495 and 1604, with wages increasing at a much lower rate. The real terms wages of agricultural and building labourers, and skilled craftsman reduced by 60% during the 16th century (Boyer, 1990). To add to this, several poor corn harvests compounded the inflationary problems. At the same time that the number of people living in poverty increased, the supply of charitable support decreased. The dissolution of the monasteries (1536–1540), religious guilds, fraternities, alms-houses and hospitals destroyed much of the institutional fabric which had provided support for the poor (Slack, 1990).

These problems prompted the passing of the *Act for the Relief of the Poor 1597* establishing a relatively simple support framework, appointing 'Overseers of the Poor', responsible for distributing money, food and other essentials to those in poverty. This Act was amended in 1601 to enable the (re)creation of alms-houses or poorhouses for those unable to care for themselves. The able-bodied poor were set to work in Houses of Industry where materials were provided for the poor to work productively, the idle poor (those who refused to work) were sent to a House of Correction or even prison, pauper children became apprentices (Irving-Clarke, 2019). Implicit in these changes is a recognition that the state has a role to play in alleviating poverty and second that accommodation is a key part of supporting the poor and vulnerable. However, Boyer (1990) contends that the Elizabethan reforms to the poor laws had more to do with preventing social unrest because of the threat of starvation.

Shock two - the Industrial Revolution (1760–1840)

The Industrial Revolution, beginning c.1760, saw the development of steam power, machinery, and ultimately factories and cities. This would bring wide-scale change to whole country, and housing policy. The Industrial Revolution led to the urbanisation of large swathes of the population, who had previously been living an agricultural existence. The forced enclosure of common land was intended to compel people into the cities to work in the developing industries (Fink, 1981). The increased use of technology and other social changes caused by the new industries, e.g. new working hours and the breakdown of local/rural communities largely dispensed with the labour of elderly males (MacNicol, 2006). People with physical disabilities were increasingly reliant on poor law services, as machinery in the new industries was designed to be operated by the able bodied Many people with disabilities ended up housed in poor law institutions (Drake, 1996).

The Royal Commission into the Operation of the Poor Law 1832 and resulting *Poor Law Amendment Act (PLAA) 1834* brought housing provision for the poor under a single, national system of administration (replacing the previous parish-based system), moving the policy paradigm to national level (Irving-Clarke, 2019). A Poor Law Commission oversaw the operation of the new system at national level with parishes were now clustered into Poor Law Unions (each run by a Poor Law Board), and union workhouses established to provide relief to the poor. Relief to the poor was provided both as 'outside relief' or financial or charitable relief provided direct to the poor or as 'inside relief', the workhouse. During the 1840s, however, the workhouse became the favoured avenue of relief to the poor; other forms of relief being discouraged rather than abolished. The poor law system of workhouses, perhaps brutal, neglectful or both continued until 1929.

Also, at this time (the 1830s), Octavia Hill originated social housing as we think of it today. Her self-stated aim was to make "lives noble, homes happy and family life good". Her moralistic and paternalistic approach was that properties should be rented to tenants at a high standard and in return, they would be expected to 'moderate' (change) their behaviour. The government also took up the mantel and mandated local authorities to demolish and replace 'slum housing' in the *Housing and Working Classes Act 1890*. This was the first time the state had played a role in the provision of general housing to a noticeable extent. A decade later the *Housing of the Working Classes Act 1900* extended the 1890 Act to areas outside London, enabling councils to buy land outside their own districts.

Policy to support the poor was not the only area affected by the Industrial Revolution. Policy relating to people with mental health problems and learning disabilities, or 'lunatics' as they were often termed also changed direction. These groups were seen as a source of major disruption in the union workhouses and factories (Murphy, 1991). Unruly and unpredictable behaviour was not only a "burden to both inmates and staff" (ibid, p. 34) but they were also unable to work. This led to workhouse guardians adding their voice to others for these individuals to be removed elsewhere. The *Lunacy Act 1845* required the building of 'lunatic asylums' in all counties and created the 'Lunacy Commission' to regulate asylums, hospitals (private, voluntary, and public) where lunatics were likely to be treated. Local authorities grasped the nettle and began to build large asylums that were often architecturally impressive monuments of civic pride (Murphy, 1991). There were growing numbers of people, detained in ever larger institutions with fewer therapeutic pretensions (Gibbons, 1988).

The Industrial Revolution opened policy windows on several interrelated fronts. Changes in demographics and patterns of work into urbanised factories meant that new housing policies had to be developed to cope with the changes. This also meant that people who could have previously been gainfully employed could no longer be so as they were unable to work in the new industries – and therefore the state had to invent new ways to house and support them, ill-judged though some of the solutions turned out to be. The Industrial Revolution also amplified several policy paths that have been hard to break. Echoes of attitudes towards people requiring government support can still be heard into today's rhetoric about people reliant on benefits, or who have mental or physical disabilities. Attitudes to the residents of Grenfell Tower are not so different to attitudes to people who lived in the workhouse. The urbanisation of work and living has also been a lasting legacy. There was a significant financial cost of building factories, offices, housing and civic architecture, e.g. town halls etc., but many have lasted to the present day – including the accompanying institutional arrangements. Although, changes made to working patterns as a result of the COVID pandemic, e.g. working from home, may yet derail the patterns of work established at this time.

Shock three – the Great War

As I have discussed previously, World War I had a significant impact upon housing policy. This event was not particularly one that was predicted yet opened policy windows that were previously closed or at most ajar. To recap, these included rent strikes in areas such as Glasgow in response to high rents caused by a shortage of urban housing for those moving to cities to support the war effort. This led to rents and mortgage rates being frozen. The housing shortage was exacerbated by soldiers returning from the front lines at the end of the war, this led to the roll out of 'Homes fit for Heroes' (Yorke, 2017) via the *Housing and Town Planning Act 1919* (also known as the Addison Act) implementing the recommendations of the Tudor-Walters report. The report had highlighted the shortage of good quality homes for men returning from the war and recommended a huge housebuilding programme as well as introducing space standards (Park, 2017).

In terms of those unable to work or fight living in asylums and workhouses, the *Mental Deficiency Act 1913* built on the *Royal Commission on Care and Control of the Feeble Minded 1908* and defined people with learning disabilities as idiots, imbeciles or the feeble minded depending upon the level of their learning disability. It also created a new group; the 'moral imbecile' for people whose learning disability made them prone to 'loose morals'. Once an individual had been detained under these terms, it was in the power of the Board of Control to block the release of anyone they considered unfit to live in the community; although, there were safeguards. But as Means et al. (2008) point out, it was all too easy to secure the admission of any person with a learning disability. It also left other groups such as the homeless vulnerable to 'certification' regardless of their mental abilities. The act established a single, unified service that took people with mental illness away from all other mainstream services (Ryan & Thomas, 1980).

The *Local Government Act 1929* made significant changes. The best poor law hospitals were now re-titled 'Public Health Hospitals' and removed from the poor law system to the remit of local health authority committees. These concentrated on those with acute conditions; the elderly were viewed as chronic cases and so stayed in the remit of the workhouses (Baggott, 2004). The remaining hospital provision and workhouses were moved to the jurisdiction of counties and county boroughs, required to establish a 'Public Assistance Committee' (PAC) with workhouses re-titled 'Public Assistance Institutions' (PAI). However, the *Poor Law Act* kept the focus of providing for the vulnerable, including the elderly, squarely with the family (Means et al., 2008).

By 1930, there were around 120,000 people held in public asylums, numbers eventually peaked in 1954 at over 148,000. Scull (1993) coined the term 'warehouses of the unwanted' to describe what he saw as the detention of people who society was keen to be rid of. There was a second major expansion of people compulsorily detained in institutions. Figures from Tredgold (1952) quoted by Malin et al. (1980) indicate a rise from 12,000 in 1920 to 90,000 by 1939. Most of these were detained in large, isolated colonies (Means et al., 2008). More positively, the *Mental Health Treatment Act 1930* changed the name of 'asylums' to Mental Health Hospitals and introduced new categories of voluntary and temporary patients. Pearce (2007) argues that this led to more people coming forward at an earlier stage of their mental health problems and led to the introduction of more innovative treatments. However, he balances this by pointing out it also led to the rapid expansion of "unpleasant and hazardous shock therapies" (p. 128) (Electro-Convulsive Therapy [ECT] and Insulin Coma Therapy [ICT]).

World War I opened a significant housing policy window as once again patterns of living were interrupted. Working age, able-bodied males were conscripted to fight abroad, and women were moved into the cities to contribute to the war effort by manufacturing munitions. Rents rocketed as demand for housing rose, whilst demand for housing rose whilst construction all but ceased. This led to rent and mortgage controls. and the Tudor-Walters report making significant recommendations which were accepted and implemented via the Addison Act. The question is, would Tudor-Walters report or Addison Act even have been possible without the war opening the policy window, as soldiers returned from the war and government concern that more homes were needed, and living standards poor, possibly leading to unrest. The role of hundreds of soldiers returning from war with physical injuries and 'shellshock' (now called Post Traumatic Stress Disorder) in driving both numbers and some improvement in the treatment of disabled people should also not be ignored.

Shock four – World War II

I have also previously discussed the dramatic effect of World War II on housing policy. Again, to recap briefly, the government was faced with the twin problems of a housing shortage (caused by bomb damage) and returning soldiers, a "baby-boom" and the resulting growth in household formation (Irving-Clarke, 2016). The Beveridge Report (1942) introduced the concept of a comprehensive state safety net to protect its citizens from the five giant evils (squalor, ignorance, want, idleness and disease) from cradle to grave. These needs were met by a large-scale council house building programme, with councils adding 'landlord' to their growing list of municipal and public health functions. The focus on a comprehensive welfare state endured for the next five decades, indeed the Conservatives fought the 1951 General Election on a manifesto that read:

> Housing is the first of the social services. It is also one of the keys to increased productivity. Work, family life, health and education are all undermined by crowded houses. Therefore, a Conservative and Unionist Government will give housing a priority second only to national defence.
>
> *Conservative Party (1951)*

Despite seeing housing as second only to national defence at the 1951 General Election, the Conservative government of Harold Macmillan (1957–1963) was the first to challenge the 'post-war consensus' by questioning the mass provision of housing by the state. Housing is often cited as 'the wobbly pillar' under the welfare state (Torgerson, 1987), because of the partial extent in which it has been considered as a responsibility of government and of social policy.

In terms of housing the vulnerable, the *Housing (Homeless Persons) Act 1977* continued the residualisation of the poorest people in council housing, local authorities now had a duty to house people considered vulnerable under the law and provide financial assistance to help them afford the rents. Council housing was thus used to discharge this duty and the introduction of compulsory rent rebates (subsequently Housing Benefit and now the housing element in Universal Credit) meant that, as stated previously, affordability was no longer a disqualifying factor; although Malpass and Victory (2010) find that this residualisation was implicit in and result of housing policy throughout the 20th century.

In a similar vein to the first war, World War II opened significant policy windows as able-bodied, working aged men were conscripted to fight abroad, and women worked in factories etc. This time however, the return of soldiers and resulting surge in household formation were coupled with a serious shortage of homes as swathes of the country had suffered bomb damage, which had not been so prevalent in the first war. This was particularly true in London, the naval ports, e.g. Portsmouth and Plymouth, and the other manufacturing bases, e.g. Coventry. Alongside the election of Margaret Thatcher in 1979, the policy window opened at this time enabled perhaps the most ideologically relevant set of policies in British history. The idea that those who had sacrificed so much in the war deserved a better country to come back to surpassed even the Homes for Heroes initiative following the first war. As previously detailed, the setting up of a comprehensive welfare state providing housing, healthcare, education, and universal welfare benefits was planned and delivered. Hundreds of homes, hospitals and other municipal services were built. The opening of this new policy window at this time is perhaps most graphically illustrated by the shock defenestration of the victorious Winston Churchill in favour of Clement Attlee's Labour who promised this comprehensive post-war reconstruction. Again, the policies implemented at this time echo into the present. The most obvious being the NHS, which has once again proved its value during the Covid-19 pandemic.

Shock five – the Oil Crisis, and the winter of discontent

The crises that opened the window for the election of Margaret Thatcher's Conservatives in 1979 had their roots long before, in the direct post-war period. The war had been won, all that remained was to win the peace. There was a moment where there looked to be agreement that the sense of common endeavour that was engendered by the war effort must not be wasted (Blond, 2010). However, instead, sectional class interests were resurrected, and the British turned in on themselves. Blond (2010) contrasts the European approach to rebuilding with the British one. Where France, Germany and others utilised government as the coordinator of the industrial effort, as a mediator between interests that would otherwise have competed, in Britain the state and industry found themselves in conflict.

For example, Germany ensured that company boards were made up of all interested parties, including representatives of the workers. This meant that when workers felt they were not sharing adequately in the profits, or where pay restraint was called for, workers had trusted representation on company boards and could be assured their interests were being robustly represented. In the UK, this was not the case as industry and workers found themselves at almost permanent loggerheads. Strikes and other industrial action were commonplace, there was conflict as workers and industry fought one another for a share of the ever-shrinking economic cake. Britain found itself unable to innovate, trade and prosper and gradually became "the sick man of Europe" (Blond, 2010, p. 94), culminating in the 'winter of discontent' in 1978/1979. Between 1950 and 1973, average GDP growth in the UK was only 2.5%, whilst the western European average was 4%; Germany and Italy both averaged 5%. These countries managed to eliminate around 40% of the productivity gap between them and the US, with France at 74%, Italy 78% and Germany at 79% of US productivity, measured by hours worked as a percentage of GDP. Exports fared little better with the UK annual export growth rate between 1950 and 1973 at only 6.9%. This can be compared to an EU average export growth rate of 12.2%, and a remarkable German export growth rate of 19.8% (Eichengreen, 2007).

According to Blond (2010), there was a "false Keynesian consensus" (p. 102) in the UK, between owners who were only interested in higher personal dividends and workers who were only interested in increasing wages for work of reducing quality and efficiency. Both retreated into their comfort zones. Management became diffident, unvisionary and lazy, choosing comfort over innovation and challenging workers. The labour unions (workers) retreated into opposing anything that seemed to challenge their narrow, sectional interests and 'class struggle'. This managerial incompetence, and worker intransigence (segueing into ideological hostility to capitalism itself), proved a potent cocktail. The British people turned to the most "radical and charismatic European politician of the post-war era" (Blonde, 2010, p. 94), Margaret Thatcher. Thatcher saw the stakes for what they were and began to restructure the economy, but in doing so, whole swathes of the population were abandoned to a life of unemployment or low wages.

Although some argue that the neo-liberalisation of welfare policy had been in train for some time (see below), the election of a transformed Conservative Party under Prime Minister Margaret Thatcher in 1979 and her adherence to the monetarist policies of Milton Friedman, which prescribed limited state intervention in a 'self-correcting' free market, drove housing policy during the 1980s and well beyond (Irving-Clarke, 2016; 2019). Although, Gough (1980) suggests that the ideological attack on the post-war welfare state was largely initiated by the Labour governments of Wilson and Callaghan. He further argues that the qualitative shifts in social policy at this time, were aimed at reasserting individualism, promoting self-reliance and family responsibility; and to ultimately dismantle the ideology of the collective social provision of the post-war era. The Thatcher governments of the 1970s and 1980s focused on setting people free from government constraints on their lives by rolling back the state, re-emphasising a commitment to market provision of housing, and a revival of the private rented sector.

Driving policy was a preference for home ownership over other tenures (see Gurney, 1999). The *Housing Act 1980* introduced the flagship 'Right to Buy' policy. Local authority that enabled to purchase their property at a discount from the market price, dependent on how long they had been a tenant. The policy proved extremely popular with voters and has never been fully repealed by successive governments of either party in England; although the *Housing Act 2004* introduced stricter regulations including a tighter time frame in which tenants could sell after purchase and with less discount to buy (King, 2006). The Scottish and Welsh governments have both recently abolished it in a bid to protect social housing stock from further reductions, and to ensure that the state can provide safe, secure accommodation to those who are unable to buy or rent their own home in the private market (Stephens, 2019; Irving-Clarke & Henderson, 2021).

Conservative governments of the 1980s and 1990s continued the doctrinaire repositioning of state-controlled housing out of the Government arena. The *Housing and Planning Act 1988* gave councils the option of transferring all or part of their housing to an RSL. The 1988 Act also allowed council tenants (as a whole) to choose to transfer their existing homes to another landlord, either a separate housing association (LSVT) or ALMO, operated by the local authority but managed autonomously.

The effects of 'Right to Buy' were manifold and widespread. It removed much of the most desirable and sought-after stock from the social housing sector, without a replacement building programme (Forrest & Murie, 1988). Since the introduction of the policy, it is estimated that over two million properties have been moved into private ownership (Manns, 2017). It served to deepen the divide between owner occupation and the social rented sector. Lund (2011, p. 142) argues that Right to Buy and other policies resulted in a continued concentration of low-income households in the social housing sector with those accessing social housing increasingly in the greatest housing and social need. Social housing was becoming more synonymous with welfare and a clear polarisation of the housing market was established. Hills argues that housing policy hardly existed anymore but survived within the welfare state, "mainly as an adjunct of social security (through Housing Benefit) or as part of wider city regeneration policies" (Hills, 1998, p. 13 quoted in Tunstall, 2015).

Housing policy (arguably) changed again with the election of New Labour under Tony Blair in 1997. This was under the banner of the 'Third Way' (Blair, 1998). A supposed route between the first (social democracy) and second (neoliberal) ways of the post-war period. Blair argued that the third way was a route between these two ideologies, whilst still being firmly on the left of British politics (Blair, 1998); third way politics has been criticised on several grounds. First, because it merely continued the polices of the previous Conservative governments of the 1980s and 1990, and second, because of its pretensions to present a route between neo-liberalism and social(ism) democracy, the third way has been criticised as being an ideological vacuum; pragmatism in place of ideology. Further, it has been asked whether the third way is really a radical programme of renewal or a clever marketing gimmick for a continuation of Thatcherism – neoliberalism with a human face (King, 2001). If nothing else, the transfer from the neoliberal policies of Thatcher and Major to the 'third way' policies of Blair and Brown was marked by a relatively smooth transition with no large exogenous shocks. Perhaps this is reflected in the perceived continuity in housing policy, and main criticisms of the third way given above?

New Labour continued the policy of seeing home ownership as the pinnacle of the housing system, although they also initiated a major investment in the social housing stock under the Decent Homes Programme (DETR, 2000), invested in housing-related support via the Supporting People programme (DTLR, 2001) and made major inroads into tackling homelessness and rough sleeping. However, the Thatcherite 'right to buy' policy continued as did LSVTs and creation of ALMOs to transfer housing stock out of direct local authority control. These measures further residualised social housing as a tenure for those on the lowest incomes (CIH, 2018), although there was recognition that welfare and housing policies needed to be better aligned (Malpass & Victory, 2010).

The oil crisis and winter of discontent opened another significant policy window, probably the most significant since 1945. The problems between workers and industry widened to breaking point. A series of strikes in the 1970s left rubbish uncollected for weeks on end and the dead unburied to note two of the most infamous examples. On top of this, power cuts were commonplace, leading to the Conservative government led by Edward Heath to bring in a three-day working week to conserve electricity.

The political environment was also particularly volatile with a general election in 1970, followed by two in 1974, the first of which saw a minority Labour government under Prime Minister, Harold Wilson, which soon collapsed. Wilson was then re-elected with an overall majority, before transitioning to the leadership of James Callaghan. By 1979, the country must surely have been crying out for some stability. They turned to the most radical leader the Conservative Party has ever seen (see above). The post-war ideological experiment was over, brought low by both ideology and events. Despite the ideological differences continuing into the 1980s, Thatcher defeated Michael Foot's Labour Party in 1983 – many would say saved by another unpredicted event, Argentina's invasion of the Falkland Islands. This made Thatcher a 'war leader' with all the cache that it brings. Her revolution continued into the 1980s and 1990s with first Neil Kinnock modernising Labour and kicking out the 'militant tendency', then John Major's surprise victory in 1992, when many thought he would lose. Thatcher's work seemed completed in 1997 with the election of New Labour under Tony Blair. Blair was unashamedly Thatcherite in many policy areas pursuing both Conservative spending plans, in his first term, and the marketisation of the NHS. None of this would have been possible, maybe even thinkable, without the tumult of the 1970s – it did herald a period of relative calm, however.

Shock six – the 2008 financial crisis

If the transition from the Conservatives to New Labour in 1997 was a largely uneventful coup (if it was a coup at all), the switch back to a more neoliberal policy approach under a Conservative and Liberal Democrat coalition was anything but. Year 2008 saw a major financial crisis, the roots of which had been in train for some time (Blond, 2010). Blond argues that the commodification of nearly everything that could be commodified, the creation of a global marketplace for capital and overleveraging of governments, banks, families, and individuals led to a situation where debt could not be sustained in an economic downturn. Governments across the west were forced to 'bail out' banks by recapitalising them. Many of these banks, which were deemed 'too big to fail' had allowed themselves to get into a situation where they did not have enough reserve capital to meet their debts in such a crisis. This recapitalisation of the banks alongside the economic downturn meant that the UK government (and others) had to take on billions of pounds in debt from the international markets. New Labour had already incurred significant public debt because of high levels of public spending. In 1997, Labour inherited national debt of under 40% of GDP. By 2003, they were running a budget surplus, adhering to Conservative spending plans (as they had promised to, to fend off accusations of profligacy). New Labour began to run increasing budgetary deficits between 2003 and 2008, during the biggest boom in GDP on record. This is not necessarily a problem in and of itself, but it did mean that the UK government was already carrying a considerable debt when it was forced to borrow billions of pounds to recapitalise the banks and sustain public spending as tax receipts crashed.

This opened a window for the incoming Conservative and Liberal Democrat coalition government to implement considerable cuts to government spending to reduce the budget deficit (the annual increase to the national debt), and eventually the national debt itself. Even the former Labour chancellor of the Exchequer, Alistair Darling, was warning there would have to be cuts to public spending worse than those made under Thatcher (Darling, 2011). Under the rubric of reducing public spending (or austerity), the coalition government embarked upon an ambitious programme of public service reform. Indeed, governments across Europe retrenched public spending under the banner of 'austerity', calling for government and individuals to live within their means (Farnsworth & Irving, 2017). This meant decreased public spending on welfare and the institution of greater conditionality in the welfare system – said to act as incentives to work (Watts & Fitzpatrick, 2018).

In terms of housing policy, the coalition government continued with the previous Conservative (and arguably New Labour) predilection for home ownership. The coalition government failed to govern the housing system, having given up some of the powers to do so via the localisation of the planning system and budget cuts (Tunstall, 2015). Further, the coalition continued the trend of shifting expenditure from object (houses) to subject (people) subsidies, this did not deliver anywhere near the levels of affordable (social) housing needed (Tunstall, 2015).

Alongside this, the coalition made significant and lasting changes to the welfare system that had profound effects on housing policy (if housing policy was distinguishable from welfare policy by this point). First, their flagship welfare policy UC attempted to simplify the benefits system by replacing six means-tested benefits, including housing benefit, and tax credits with a single monthly payment. UC, alongside other welfare reform measures, made many households worse off than under the system it replaced (Schmuecker, 2017). This was exacerbated by the freeze of Local Housing Allowance (LHA) rates (CIH, 2016).

Increased conditionality in the system included the use of sanctions (non-payment of benefits) for those who did not meet the requirements set out for them in their agreement with DWP. In an echo of Octavia Hill's housing system, the introduction of a more punitive system was intended to push people towards behaving in the way the state favoured – this did not occur. UC was found to diminish work incentives (Commonweal Housing, 2017), meeting the conditions set within UC proved problematic and a bureaucratic nightmare for those working multiple jobs, irregular hours etc. – leading to further sanctions (Commonweal Housing, 2017; Wright et al., 2018). Johnsen et al. (2018) found that sanctions did little to motivate homeless people to seek or prepare for work, caused considerable distress, and might have pushed some extremely vulnerable people out of the social safety net altogether.

The 'removal of the spare room subsidy' or 'bedroom tax' (as it has become known) was an attempt to deal with the problem of households with more rooms in their homes than they needed, and to deal with what they perceived as a disparity between social and private tenants. Social tenants could have the full rent paid on their homes regardless of whether they needed a home of a given size, private tenants were given an allowance according to their household size and had to make up any shortfall if they wished to rent a property larger than their assessed need. People over the eligible age for pension credit were, however, exempted (see Gibb, 2015 or Moffatt et al., 2015 for fuller assessments). The impact of the bedroom tax has been harmful and has not addressed its own policy goals (Gibb, 2015). Households have less space than they often need; this particularly affects lone parents with child access arrangements and people with disabilities (Wilcox et al., 2017).

Discretionary Housing Payments (DHPs) have been used as a panacea for mitigating cuts to housing benefits and the 'bedroom tax'. This causes several issues. First, DHPs now only provide a temporary fix for loss of income that was previously a statutory entitlement and furthermore often places further conditionality on that assistance. Second, Meers (2015) states that the courts have found some of these reductions, notably the bedroom tax, to be unlawful; but allowed them on the basis that DHPs provide an adequate system of mitigation.

The overall benefit cap, originally set at £26,000, the level of median household earnings reflected the government's conviction that no household on benefits should receive more than this on equity grounds. This has since been decreased to £20,000 outside and £23,000 inside London – these are completely arbitrary figures. Research by the Chartered Institute Housing (CIH, 2017) found that the effect of the cap was disproportionately felt by larger families and was causing real hardship and in the worst cases homelessness. Children have been disproportionately affected despite having no control over their circumstances. The cap has limited the ability of larger families to meet rents in some areas. Fitzpatrick et al. (2018) found that four out of five households with four children faced at least a £50 shortfall in their rent, rising to over 90% for five child households (CIH, 2016). CIH also found that the cap is not making work the best route out of poverty (as promised), is manifestly unfair in some cases, e.g. where children have been fostered and impacts heavily upon households with no means of escape; for example, where there are very young children (CIH, 2017).

After a period of relative calm between at least 1997 (possibly earlier) and 2008, events once again gravitated towards a change in ideological direction. Despite famously positioning himself as the 'heir to Blair', many would say David Cameron (and George Osbourne) were heirs to Thatcher. They embarked on one of the biggest shake ups of the welfare system since the 1970s, rolling several means tested benefits into Universal Credit, limiting housing payments in both the private and social rented sectors. They also broke the agreed rent increase formula for social landlords, instead mandating a 1% rent reduction for five consecutive years. This continues to impact to the present day as the impacts are cumulative, even after rents began to increase again. Once again, global events enabled a sharp change in direction by opening window for policy change. The Conservatives made the most of the opportunity restoking old fears that Labour were profligate spenders and had bankrupted the country, whilst convincing many that the solution had to be spending cuts and changes to the welfare system.

Shock seven – the Grenfell Tower Fire

On 14 June 2017, a fire broke out in the Grenfell Tower block of flats in the Royal Borough of Kensington and Chelsea, London. The fire caused the death of 72 people, including two who died later in hospital, and injured more than 70 others. The subsequent inquiry revealed serious shortcomings in the fire safety architecture. Fire retardant doors were missing or in poor repair with missing tumescent strips, fire stops in internal access chutes were missing. The building had a 'stay put' policy in case of fire, designed so that the fire doors and stops prevent spread long enough for the fire service to respond and evacuate the building.

The building had been covered in flammable cladding that gave off poisonous gases as it burned. The tests that had been used on the cladding for fire retardancy have been proven wholly inadequate. Even if the internal fire doors and stops has been in place, they would have been rendered largely useless as the fire spread up the outside of the building via the external cladding. On top of this, the fire alarm proved wholly inadequate, with many residents who escaped saying they had not heard it. By the time many of the residents decided to evacuate, the building was an inferno, filled with toxic gas. It was a perfect storm.

The fire at Grenfell Tower led to a public inquiry (that is ongoing at the time of writing); a government commissioned independent review of building regulations and fire safety chaired by Dame Judith Hackitt, culminating in the Hackitt Report [Cm9607]), and both a green paper (MHCLG, 2018) and a white paper (MHCLG, 2020). The Hackitt review found multiple failings in building safety including a lack of an audit trail as to whether essential safety work had been carried out, and both fire doors and the cladding on the building had failed under testing but still installed in/on the block (Cm9607). As well as these more technical failures, Hackitt was also critical of the systemic and cultural failings surrounding the fire. She highlighted the main issues underpinning the system failure as:

- **Ignorance** – regulations and guidance were not always read by those needing to do so and when they do the guidance may be misunderstood or misinterpreted.
- **Indifference** – the primary motivation was to build as quickly and cheaply as possible at the cost of quality. When concerns were raised, they were often ignored; some involved in the construction failed to prioritise building safety, instead using the ambiguity of the regulations to game the system.
- **Lack of clarity on roles and responsibilities** – fragmentation of the construction industry led to ambiguity over where accountability lies, precluding ownership of responsibility.
- **Inadequate regulatory oversight and enforcement** – the size and/or complexity of a scheme does not seem to affect oversight. Where enforcement was necessary, it was not always pursued, and penalties were perceived as ineffective.

The fire at Grenfell Tower has proven a pivotal moment in housing policy terms. Unlike some of the previous changes discussed, there was no set of policy prescriptions on the shelf, or in the 'primeval soup' ready to go, to address the causes of the fire, or the failures in building safety and tenant engagement. Instead, the government had to embark on lengthy review and inquiries into the causes with a view to changing building safety policy and the policy approach in general.

First a Social Housing Green Paper (MHCLG, 2018) and then the Social Housing White Paper (SHWP) (MHCLG, 2020), picked up on these themes introducing a new pro-active regulation regime, measures on professionalisation and stigma against social housing tenants, building safety and improving tenant engagement and voice (including complaint handling) (see CIH, 2020). Alongside the white paper, the government also passed the *Building Safety Act 2022*, giving residents of high-rise blocks more say in the management of them and the ability to raise safety concerns with the owners and managers of buildings, who are under a duty to listen; where residents feel they are being ignored, they will be able to raise their concerns directly with the Building Safety Regulator. This will apply retrospectively to buildings constructed up to 15 years prior to the Act. Building owners are now required to pro-actively manage safety risks, with clear accountability for safety during design, construction and completion. There will need to be a 'golden thread' file that considers and records safety at every stage of the building's lifecycle; and proportionate measures will need to be in place to manage safety risks. Failure to do this may result in criminal sanctions.

This is a good example of how an unexpected, albeit tragic, event can change the policy climate and even draw governments out of their ideological comfort zone. It's probably fair to say that the Cameron/Osbourne governments were within this zone when they were reducing and transforming the welfare system, many would argue, along ideological lines (Farnsworth & Irving, 2017). Teresa May, despite her speech on the steps of Downing Street, did not show any major interest in reversing the changes. Indeed, Universal Credit is a good example of a very difficult policy to reverse. Who wants to redesign the welfare system again, so soon? May was, however, largely out of her depth with another of Cameron's legacy projects, leaving the EU. Again, despite her strong words, she never wanted to do so and didn't have the confidence, or the parliamentary votes, to effect, her version of it. In fact, it took Boris Johnson, who's time many had thought had passed, to give effect to referendum result. For good or for ill, the result has been respected.

The Grenfell Tower fire did not only open a policy window for change to be affected, but it also forced the change. To do nothing, or act inadequately, in the face of such a tragedy and the public outcry, would sure have been political suicide. Despite the fact that ramping up the powers of the state and imposing regulation is not exactly in the Conservative comfort zone, the government, nevertheless, embarked upon a wholesale redesign of the building safety regulations, created one to regulator and considerably increased the powers of another and the Housing Ombudsman Service, which deals with complaints from tenants. Has a government ever been forced to act against its ideological instincts in such a manner? The changes wrought in 1945 and 1979, respectively, were at least driven by the public's need and desire for the change the party proposed. This is of a different order entirely and goes to show that events can sometimes be powerful, more powerful than ideology. The policy window opened by the Grenfell Tower fire remains open to some degree.

Shock eight – COVID-19

In the early months of 2020, the World Health Organization (WHO) declared a global pandemic of a novel coronavirus SARS-COVID 2 (COVID-19). It was first identified in Wuhan, China, in late 2019, and then subsequently spreading across the world causing serious illness and millions of deaths globally. In the UK (as in many other countries), the pandemic led to several 'lockdowns' (albeit under the control of devolved governments). Like the previous discussion, the Conservative UK government is instinctively uncomfortable with high government spending, indeed rolling back the frontiers of the state has been Conservative dogma since at least 1979. However, to prevent the spread of COVID-19 and protect the NHS from being overwhelmed the government embarked on several extremely costly policies. Chiefly, the furlough scheme paid for staff across most sectors of the workforce to not go to their jobs with the state paying 80% of their wages, some employers topped this up to 90% or even 100%. People who are self-employed and met the terms of the scheme were also able to claim money from the state to enable them to refrain from work. The NHS was given massive increases in funding to pay for the increase in services that were required. An ideologically frugal Conservative Party was forced into high spending policies that it would surely have baulked at, at any other time.

The pandemic also exposed deep divisions in society. Many white-collar, office workers were able to work from home or were furloughed. On the other hand, many people in blue-collar, manual and lower-paid work were either furloughed on 80% of their wages – precarious for people who might already have been struggling on their full wage or, they had to continue to go out to work as they do jobs that cannot be carried out from home or online. The negative effects of the pandemic have been concentrated in the lower end of the labour market (Brewer, 2020). For those still going out to work, it exposed them to a much higher risk of contracting COVID-19, having to self-isolate and therefore receive only statutory sick pay (Fancourt & Steptoe, 2021).

For some, who could work from home, the pandemic (further) exposed the inadequacy of their home environment (Brown et al., 2020; National Energy Action, 2020), first as a workplace as well as a home, and second as an environment where they were expected to spend 23 h a day (people were only allowed outside their homes to shop for essential and exercise). The incidence of domestic abuse increased, particularly around coercive control, as perpetrators lost control of other areas of life and women and children lost the respite of being able to leave the home (Dyakova et al., 2021). Children and young people also suffered as schools were forced to switch to online learning, and they were separated from their classmates and missed out on the socialisation that school offers. Opportunities for play were also curtailed as children were also forced to spend most of their time in the home (Children's Commissioner, 2020; Rosenthal et al., 2020).

In terms of homelessness policy, there has been more positive impacts, although not universally so. Lewer et al. (2020) found that (in England), people experiencing homelessness were generally protected by general policy initiatives, e.g. infection controls in hostels, closing dormitory style accommodation and the broader 'Everyone In' initiative. This was another high-spending policy initiative that paid for people and households who would otherwise sleep rough, to stay in self-contained accommodation, often in the hotels that had been forced to close by lockdown rules.

Subsequently, the aftermath of the pandemic has been exacerbated by the impact of a 'cost of living' crisis caused in part by the inflationary effects of the pandemic spending described above and the invasion of Ukraine by Russia. Ukraine being the source of much of Europe's wheat and other staple foods.

This is another example of events not only opening a window for change but also creating the moral and social imperative for that change. As with Grenfell, doing nothing in the face of the pandemic was not a viable option and would have had catastrophic consequences for the government had they not acted in the way that they did. Even then, there are ongoing consequences with an inquiry to look into the pandemic response and the fallout from 'partygate' which may yet see Boris Johnson defenestrated.

Conclusion

I argued in Chapter 2 on the state, the British 'first past the post' electoral system is designed to elect governments with a strong mandate and in a strong position to implement their manifesto. This is frequently cited as one of the strengths of this system. This means that, in theory at least, once elected, a government can do as it wishes. The 'whipping' system exists to discipline MPs into voting with the executive's wishes. However, as I also argue, there are also significant institutional constraints on what governments can do. Some of these constraints are uncodified, such as conventions that should not be breached. One such convention is calling of general elections. There is no piece of legislation (any longer) that compels the Prime Minister to call an election at a given time, or indeed, ever. However, there is a *convention* that no longer than five years should pass without one. Likewise, were the sovereign to refuse Royal Assent, this would provoke a constitutional crisis as a direct challenge to the authority of the elected house. There are also codified limits in place. For example, the *Parliament Act 1911* prevents the House of Lords from preventing government legislation passing into law, because of the supposed primacy of the elected house. Were England or the UK ever to move to an elected upper house (or Senate), this house would also be able to claim the legitimacy and power that comes from this. We would then have to work out how to negotiate this impasse. Likewise, were there ever to be an elected Head of State, they too would the legitimacy of elected office. The courts too, provide a check on the executive and government as I described in Chapter 2. Both elements of the unwritten constitution and written elements place limits on what the government can do. A key role of the courts is to ensure that the government does not overreach its power (or act in a way that is *ultra vires*).

In this chapter, I have tried to show some of the less formal limits on what governments are able to do. Some of these limits are practically invisible to all but the most intimate observers of policy making and implementation. Previous and current policies, institutions and path dependencies all limit the ways that governments may act. There are also both major and minor events that create 'ripples' in policy time that may open or close policy windows, enabling the implementation if policies that have been in the 'primordial soup' for some years, waiting for their time. Events may also close off some, perfectly viable, policy options.

In the narrative constructed above, I have established significant events exogenous to policy process that have significantly altered the course of housing policy, and wider social policy. Most were largely unexpected until they actually happened. These events have opened policy windows (Kingdon, 1995) and enabled some of the large-scale policy changes seen over the centuries.

References

Allen, C & Imrie, R (eds) (2010) *The Knowledge Business: The Commodification of Urban and Housing Research*, Routledge, Abingdon.

Armstrong, F & Alsop, A (2010) Co-production can contribute to research impact in the social sciences, *Public Money and Management* 30(4) pp.208–210.

Baggott, R (2004) *Health and Health Care in Britain* (Third Edition), Palgrave MacMillan, Basingstoke.

Balchin, P (1995) *Housing Policy: An Introduction* (Third Edition), Routledge, Abingdon.

Bengtsson, B & Ruonavaara, H (2010) Introduction to the special issue: Path dependence in housing, *Housing, Theory and Society* 27(3) pp.193–203.

Beveridge Report (Cm6404) (1942) *Social Insurance and Allied Services*, HMSO, London.

Blair, T (1998) *The Third Way: New Politics for the New Century*, Fabian Society, London.

Blond, P (2010) *Red Tory: How Left and Right Have Broken Britain and How We Can Fix It*, Faber & Faber, London.

Blunkett, D (2000) Influence or irrelevance: Can social research improve government? *Research Intelligence, BERA* 71 pp.12–21.

Bobbio, N (1994) *In Praise of Meekness and Other Writings*, Linea d'Ombra, Milan.

Boyer, G.R. (1990) *An Economic History of the English Poor Law, 1750–1850*, Cambridge University Press, Cambridge.

Brewer, M, Corlett, A, Handscomb, K, McCurdy, C & Tomlinson, D (2020). *Living Standards Audit 2020*. Available at: https://www.resolutionfoundation.org/app/uploads/2020/07/living-standards-audit.pdf. (Accessed 5 May 2021).

Brown, P, Newton, D, Armitage, R & Monchuk, L (2020). *Lockdown. Rundown, Breakdown: The COVID-19 Lockdown and the Impact of Poor-Quality Housing on Occupants in the North of England*. Available at: https://www.flipsnack.com/NorthernHousingConsortium/lockdown-rundown-breakdown-1hpf1x7e47/full-view.html. (Accessed 19 February 2021).

Cabinet Office (1999) *Modernising Government*, TSO, London.

Cabinet Office (2000) *Wiring it Up: Whitehall's Management of Cross-cutting Policies and Services: A Performance and Innovation Unit Report*. Cabinet Office, London.

Children's Commissioner (2020). *No Way Out: Children Stuck in BandBs during Lockdown*. [Online]. Available at: https://www.childrenscommissioner.gov.uk/wp-content/uploads/2020/08/cco-no-way-out.pdf. (Accessed 19 February 2021).

CIH (2016) *Mind the Gap: The Growing Shortfall Between Private Rents and Help with Housing Costs*, CIH, Coventry.

CIH (2017) *Feeling the Pinch: The Lowered Benefit Cap One Year On…*, CIH, Coventry.

CIH (2018) *Rethinking Social Housing*, CIH, Coventry.

CIH (2020) *What You Need to Know About the Social Housing White Paper*, CIH, Coventry.

Darling, A (2011) *Back from the brink: 1000 days at Number 11*, Atlantic Books, London.

DETR (2000) *Quality and Choice: A Decent Home for All: The Housing Green Paper*, DETR, London.

Drake, R (1996) *A Critique of the Role of Traditional Charities* in L Barton (ed), *Disability and Society: Emerging Issues and Insights*, Longman, Harlow, pp.147–166.

Drake, R (2001) *The Principles of Social Policy*, Palgrave, Basingstoke.

DTLR (2001) *Supporting People: Policy into Practice*, DTLR, London.

Dyakova, M, Couzens, L, Allen, J, Van Eimeren, M, Stielke, A, Cotter-Roberts, A, Kadel, R, Bainham, B, Ashton, K, Stewart, D, Hughes, K & Bellis, M.A. (2021). *Placing Health Equity at the Heart of the COVID-19 Sustainable Response and Recovery: Building Prosperous Lives for All in Wales*. [Online]. Available at: https://phw.nhs.wales/news/placing-health-equity-at-the-heart-of-coronavirus-recovery-for-building-a-sustainable-future-for-wales/placing-health-equity-at-the-heart-of-the-covid-19-sustainable-response-and-recovery-building-prosperous-lives-for-all-in-wales/. (Accessed 23 March 2021).

Eichengreen, B (2007) *The European Economy since 1945: Coordinated Capitalism and Beyond*, Princeton University Press, Princeton NJ.

Fancourt, D & Steptoe, A (2021). *Trajectories of Anxiety and Depressive Symptoms during Enforced Isolation Due to COVID-19 in England: A Longitudinal Study.* [Online]. Available at: https://www.thelancet.com/journals/lanpsy/article/PIIS2215-0366(20)30482-X/fulltext. (Accessed: 10 May 2021).

Farnsworth, K & Irving, Z (eds) (2017) *Social Policy in Times of Austerity: Global Economics and the New Politics of Welfare*, Policy Press, Bristol.

Fink, H (1981) *Social Philosophy*, Methuen, London.

Fitzpatrick, S, Bramley, G, Sosenko, F, Blenkinsop, J, Wood, J, Johnsen, S, Littlwood, M, & Watts, B (2018) *Destitution in the UK 2018*, Joseph Rowntree Foundation, York.

Forrest, R & Murie, A (1988) *Selling the Welfare State: The Privatisation of Public Housing*, Routledge, Abingdon.

Gibb, K (2015) The multiple policy failures of the UK bedroom tax, *International Journal of Housing Policy* 15(2) pp.148–166.

Gibbons, J (1998) Residential care for mentally ill adults in I Sinclair (ed), *Residential Care: The Research Reviewed*, HMSO, London pp.157–197.

Giddens, A (1998) *The Third Way: The Renewal of Social Democracy*, Polity, Cambridge.

Gough, I (1980) Thatcherism and welfare state: Britain is experiencing the most far-reaching experiment in 'new right' politics in the western world, *Marxism Today* pp.7–12.

Gurney, C.M. (1999) Pride and prejudice: discourses of normalisation in public and private accounts of home ownership, *Housing Studies* 14(2) pp.163–183.

Hills, J (1998) *ESRC Research Centre for Analysis of Social Exclusion: Annual Report 1998/9*, ESRC, London.

Irving-Clarke, Y (2016) *Supporting People: How Did We Get Here and What Does it Mean for the Future?* Unpublished PhD Thesis, De Montfort University, Leicester.

Irving-Clarke, Y (2019) *Supported Housing: Past, Present, and Future*, Routledge, Abingdon.

Irving-Clarke, Y & Henderson, K (2021) *Housing and Domestic Abuse: Policy and Practice*, Routledge, Abingdon.

Johnsen, S, Fitzpatrick, S, & Watts, B (2018) Homelessness and social control: A typology, *Housing Studies* 33(7) pp.1106–1126.

King, P (2001) *Contemporary Housing Policy and the Politics of the Impossible: Zizek, Kant and the Third Way*, De Montfort University, Leicester.

King, P (2006) *Choice and the End of Social Housing*, IEA, London.

King, P (2010) Using big ideas: The application of political philosophy in housing research, *Housing Theory and Society* 28(2) pp.109–122.

Kingdon J (1995) *Agendas, Alternatives and Public Policies* (Second Edition), Harper Collins College Publishers, New York.

Lewer, D. Braithwaite, I. Bullock, M. Eyre, M.T. White, P.J. & Aldridge, R.W. (2020). COVID-19 among people experiencing homelessness in England: A modelling study. *Lancet Respiratory Medicine.* (in press, corrected proof). Available at https://www.thelancet.com/journals/lanres/article/PIIS2213-2600(20)30396-9/fulltext. (Accessed: 5 May 2021).

Lund, B (2011) *Understanding Housing Policy*, Policy Press, Bristol.

MacNicol, J (2006) *Age, Discrimination: An Historical and Contemporary Analysis*, Cambridge University Press, Cambridge.

Malin, N et al. (1980) *Services for the Mentally Handicapped in Britain*, Croom Helm, London.

Malpass, P (1990) *Reshaping Housing Policy: Subsidies, Rents and Residualisation*, Routledge, Abingdon.

Malpass, P (2011) Path dependence and the measurement of change in housing policy, *Housing Theory and Society* 28(4) pp.305–319.

Malpass, P & Murie, A (1999) *Housing Policy and Practice* (Fifth Edition), MacMillan, Basingstoke.

Malpass, P & Victory, C (2010) The modernisation of Social Housing in England, *International Journal of Housing Policy* 10(1) pp.3–18.

Manns, J (2017) The housing crisis will only get worse until England scraps the right to buy, *The Guardian*, October 11, 2017.

Marsh, D & Rhodes, R.A.W (1992) *Policy Networks in British Government*, Clarendon Press, Oxford.

Meagher, J & Lyall, C (2013) The invisible made visible: using impact evaluations to illuminate and inform the role of the knowledge intermediaries, *Evidence and Policy* 9(3) pp.409–418.

Means, R, Richards, S & Smith, R (2008) *Community Care* (Fourth Edition), Red Globe Press, London.

Meers, J (2015) Panacean payments: The role of discretionary housing payments in the welfare reform agenda, *Journal of Social* Security Law 22(3) pp.115–129.

MHCLG (2018) *A New Deal for Social Housing*, MHCLG, London.

MHCLG (2020) *The Charter for Social Housing Residents: Social Housing White Paper*, MHCLG, London.

Moffatt, S, Lawson, S, Patterson, R, Holding, E, Dennison, S & Bown, J (2015) A qualitative study of the impact of the UK 'bedroom tax', *Journal of Public Health* 38(2) pp.197–205.

Mullins, D & Murie, A (2006) *Housing Policy in the UK*, Palgrave MacMillan, Basingstoke.

Murphy, E (1991) *After the Asylums: Community Care for People with Mental Illness*, Faber & Faber, London.

National Energy Action. (2020) *UK Fuel Poverty Monitor.* [Online]. Available at: https://www.nea.org.uk/researchpolicy/ukfpm-2020/. (Accessed 5 May 2021).

Nutley, S.M., Walter, I & Davies, H.T.O. (2007) *Using Evidence: How Research Can Inform Public Services*, Policy Press, Bristol.

Park, J (2017) *One Hundred Years of Space Standards: What Now?* Levitt Bernstein, London.

Pearce, D (2007) Family, gender and class in psychiatric patient care during the 1930s: The 1930 mental treatment act and the Devon mental hospital in P Dale & J Melling (eds), *Mental Illness and Learning Disability Since 1850: Finding a Place for Mental Disorder in the UK*, Routledge, Abingdon, pp.122–130.

Pierson, P (2000) Increasing returns, path dependence, and the study of politics, *The American Political Science Review* 94 pp.251–267.

Pollitt, C (2008) *Time, Policy, Management: Governing with the Past*, Oxford University Press, Oxford.

Rosenthal, D.M, Ucci, M, Heys, M, Hayward, A & Lakhanpaul, M (2020). Impacts of Covid-19 on vulnerable children in temporary accommodation in the UK. [Online]. Available at: https://www.thelancet.com/journals/lanpub/article/PIIS2468-2667(20)30080-3/fulltext. (Accessed 25 March 2021).

Ryan, J & Thomas, F (1980) *The Politics of Mental Handicap*, Penguin, London.

Sarewitz, D & Pielke, R.A (2007). The neglected heart of science policy: Reconciling supply of and demand for science, *Environmental Science & Policy* 10, 5–16. http://dx.doi.org/10.1016/j.envsci.2006.10.001.

Schmuecker, K (2017) *Universal Credit: A Joseph Rowntree Foundation briefing*, Joseph Rowntree Foundation, York.

Scull, A (1993) *The Most Solitary of Afflictions: Madness and Society in Britain 1700-1900*, Yale University Press, New Haven CT.

Slack, P (1990) *The English Poor Law, 1531–1782*, MacMillan, London.

Smith, K (2010) Research, policy and funding–academic treadmills and the squeeze on intellectual spaces, *British Journal of Sociology* 61 pp.176–195.

Stephens, M (2019) Social rented housing in the (dis) United Kingdom: Can different social housing regime types exist within the same nation state? *Urban Research and Practice* 12(1) pp.38–60.

Streeck, W & Thelen, K (2005). Introduction: institutional change in advanced political economies in W Streeck & K Thelen (eds), *Beyond Continuity: Institutional Change in Advanced Political Economies,* Oxford University Press, Oxford, pp.1–39.

Torgerson, U (1987) Housing: The wobbly pillar under the welfare state, *Scandinavian Housing and Planning Research* 4(1) pp.116–126.

Tredgold, A (1952) *A Textbook on Mental Deficiency,* Balliere, Tindall and Cox, London.

Tunstall, R (2015) *The Coalition's Record on Housing: Policy, Spending, and Outcomes 2010–2015,* Centre for Housing Policy, York.

Watts, B & Fitzpatrick, S (2018) *Welfare Conditionality,* Routledge, Abingdon.

Wilcox, S et al. (2017) *UK Housing Review 2017,* CIH, Coventry.

Wright, S, Dwyer, P, Jones, K, McNeill, J, Scullion, L & Stewart A.B.R. (2018) *Final Findings: Universal Credit,* Dept. of Social Policy & Social Work, York.

Yorke, T (2017) *Homes Fit for Heroes: The Aftermath of the First World War 1918–1939,* Countryside Books, Newbury.

EPILOGUE – CONCLUDING COMMENTS

Chapter 1 looks at ideology. These are the (very) big ideas that underpin our views of pretty much everything else in the world. When we are navigating our way in the world, we must start with some base assumptions about it to provide us with a framework of how to act towards the other people, animals, and objects around us. For example, most of us operate on the basis that all people are sacred and there are certain things that cannot be done. In the UK (and some other places), we accept that life is sacred and no one person generally has the right to end the life of another. There are circumstances when this can be set aside, for example, to protect our own life or the life of others, but this is still grounded in the belief of the sanctity of life and its protection. Likewise, in wars and conflicts, the taking of life is deemed acceptable, if regrettable in defence of a cause. One of the most profound observations I have ever heard on this came from a former ambassador and negotiator who noted that 'no one thinks they are in the wrong'. Even the Nazis thought they were fighting a noble cause rooted in their world view – the ideology of Nazism. Both sides justified the taking of life in defence of their views. Fortunately, the allied forces prevailed.

In other places, such as some states in the US, the death penalty is still retained as the ultimate deterrent punishment for particularly heinous crimes. This is also rooted in the 'eye for an eye' philosophy of the Christian bible (Exodus 21:23–27) which sets out that the punishment for a crime should equal the impact on the victim. This principle is retained in the English common law; plaintiffs are entitled to compensation equivalent to their losses. Here we come to Chapter 4 on social justice. One of the great geniuses of the common law system is that it works on a system of precedent, looking to past cases for examples of what should happen in present cases. There is a mass of jurisprudence in this area and there is an ongoing conversation in law regarding ensuring the punishments fit the crime.

DOI: 10.1201/9781003298212-8

Ideology also underpins out views about housing. As an essential component of life that is provided for many by the state, the distribution of housing has been one of the key political arguments for many decades. As I discuss earlier, conservatives generally prefer the (relatively) free market as the primary method for distributing housing (and other commodities). This is partly because the market will enable those with the greatest resources to access the best housing and this therefore acts as an incentive to work hard and accumulate capital. However, conservatives are also inherently suspicious of the state and the state having too much power over people's lives. They therefore view private property (Chapter 5) and housing, in particular, as a bulwark against the overmighty state. After all, if the state controls your ability to live in your home, what can't they make you do? This is an extreme example, but you get the point. Controlling a person's home and whether or not they are allowed to have one gives a significant amount of leverage. Indeed, one of the frequent attack points from conservatives is that state-provided housing does act as a 'bribe' come voting time.

Those on the 'left' of politics, however, are more concerned with social justice as an issue. Whilst conservatives might argue that it is just that the most productive have access to the best housing, those on the left argue that there are larger issues at play. As good quality, affordable housing is *so* important to human thriving, the left argue that that state should provide housing for people, using funds derived from taxation of those deemed wealthy enough to be able to pay. Their ideology is much more based in the idea of fairness and equality and ensuring that those of less means are still able to have a certain standard of living. This, of course, opens up questions about who gets what, and this the basis of the discussion of social justice. This is partly a question of 'need', how big a home does a given household need? Should this household have access to a given size of home based purely on the number of people to be housed? What happens if a family is living in a house of appropriate size, but they choose to have more children? Are they entitled to a larger home, or have they undertaken a deliberate act that has caused them to be overcrowded? Is it fair to offer them a larger home when there are households living in temporary accommodation or in cars or caravans who would derive greater benefit from the same home? The left's ideology, based on tenets of equality and fairness, would state that people in similar circumstances should be treated in a similar way, although this does not remove the issue of allocating scarce resources.

Chapter 2 looks at the development and role of the state. Borrowing from the discussion above, there are contrasting views on the role of the state. Conservatives are generally in favour of a more limited state that intrudes as little as possible in people's lives. Whereas, those on the left favour a more interventionist state that plays a key role in sharing the resources that society generates across that society. Their view is that, left to their own devices, the owners of capital will accumulate all the resources whilst paying workers, who generate the wealth, as little as possible. And there is something to this idea. There is indeed a balance between left and right that needs to be struck. Again, the location of this 'point' is a subject of some argument.

The state also plays a key role in other areas of housing policy. The state provides an authority figure, via the law and the courts, and a framework that supports the market (Chapter 3) and private property (Chapter 5). If we think back the ideas of Thomas Hobbes in Leviathan about what exists before the state, he describes this as a 'state of war', in this state, there are no rules, everyone is out for themselves and brutality reigns. Hobbes', and for that matter Locke's, contention is that the people come together and hand over powers to a sovereign individual or body to enforce the law and prevent injustice. One of the role of the state then is to provide a common framework, under the law, that enables the keeping of justly gained private property and provides restitution where this is violated. The state, via the courts, also provides a mediating body that can decide when the law or constitution has been infringed and award appropriate compensation. Finally, the state provides a method by which the government may violate the normal rules of ownership for a greater good. For example, there are occasions when the government wants to undertake significant infrastructure projects (HS2 springs to mind) and wishes to compulsorily purchase land to do so. In these cases, the courts may be called upon to decide if the government is acting within its powers and the rights of the property owner are being properly weighed against the benefit provided by the project.

There is also a key role for the state in providing a measure of stability for property rights and relatedly, the market, to exist and operate properly. Again, looking back at Hobbes, the state needs to provide stability by recognising and enforcing property rights. In the UK, for example, this takes the form of 'deeds of ownership' in the case of housing, papers that legally register one's ownership of a plot of land. The government also runs a land registry to keep these records and mediate in the case of disputes. In the case of smaller property items, most sellers provide invoices or receipts when you purchase something which you can use to prove ownership if challenged. Cars, for example, are required to have a unique record number, and the car owner has a V5 logbook that registers them as the owner of that vehicle. These measures, amongst others, enable ownership to be established and the state to provide mediation and restitution in cases where property rights have been violated. The state may also punish those who violate property rights by stealing or damaging other people's property.

This is important as these measures enable and maintain a stable basis upon which we can own and trade things as we wish, within some limits. This is the other role of the state that I discuss. By recognising property rights and documents that prove ownership, the state provides a basis for a market in commodities where two (or more) parties can agree on a price (value) for something and be assured that this agreement will be respected and enforced by the state. The state also provides and underwrites currency that can be used anywhere in the country as a means of exchange. This removes the need for barter, the exchange of goods of equivalent value, by providing a token with a guaranteed value that can be stored away for future use, removing the immediacy of barter trading. This also enabled the institution of a banking system to store money and facilitate exchange. The state also backs these banks via a 'central bank' and becomes the 'lender of last resort' as we saw in the example of the 2008 banking crisis in Chapter 3 on markets.

Chapter 4 looks at property rights, and as I have already said, the importance of these is maintaining a stable society via the recognition of ownership. Linking back to Chapter 1 on ideology, there are different views on property rights between the political right and left. As I have already said, the right tends to view property rights as vital, not only in terms of stability, but they also view ownership of private property, particularly housing, as a bulwark against the state. This is also closely linked to family and the ability to raise children with the values that one sees an important and avoiding the interference of the state in the process. The left, on the other hand, is much less attached to property rights, instead viewing equality and fairness as more important. The left is much more willing for and supportive of, the state in removing property from the wealthy and using this to ensure that those with lesser means are well provided for by the state – including via council or social housing.

In the final chapter, I discuss policy change and some of the factors that contribute to change or indeed, constrain it. In the English electoral system, winning political party has an overall majority of seats in the House of Commons, which has primacy over the Lords and faces no opposition from the Sovereign as a symbolic head of state only. There is often sometimes a thought then that Parliament is little more than a rubber stamp for whatever the government wished to do. However, this is far from the case. There are myriad factors that limit the options open to the government. Past policy, for example, can limit what is viable in the present, even closing off some policy options completely. Politicians are, of course, also highly sensitive to public opinion; reliant on it, as they are, for their continuing presence in Parliament. There are also precedents and conventions that limit government action by setting the rules of the game, sometimes informally and almost invisibly.

In this book, I have tried to set out some of the philosophical and conceptual ideas that underpin our policies in the present. Some of these reach back to antiquity and have proven so powerful that they continue to influence our polity and policy to this day. My hope is that this book fills a gap in the literature and is of use and interest to both philosophy and housing scholars and practitioners alike.

INDEX